Engineering and Technology for Healthcare

Engineering and Technology for Healthcare

Edited by

Muhammad Ali Imran
University of Glasgow

Rami Ghannam
University of Glasgow

Qammer H. Abbasi
University of Glasgow

WILEY

This edition first published 2021

Registered Office(s)
John Wiley & Sons, Inc., 111 River Street, Hoboken, NJ 07030, USA
John Wiley & Sons Ltd, The Atrium, Southern Gate, Chichester, West Sussex, PO19 8SQ, UK

Editorial Office
The Atrium, Southern Gate, Chichester, West Sussex, PO19 8SQ, UK

For details of our global editorial offices, customer services, and more information about Wiley products visit us at www.wiley.com.

Wiley also publishes its books in a variety of electronic formats and by print-on-demand. Some content that appears in standard print versions of this book may not be available in other formats.

Library of Congress Cataloging-in-Publication Data

Names: Imran, Muhammad Ali, editor. | Ghannam, Rami, editor. | Abbasi, Qammer H., editor.
Title: Engineering and technology for healthcare / [edited by] Muhammad Ali Imran, Rami Ghannam, Qammer H. Abbasi.
Description: Hoboken, NJ : Wiley-IEEE, 2020. | Includes bibliographical references and index.
Identifiers: LCCN 2020025470 (print) | LCCN 2020025471 (ebook) | ISBN 9781119644248 (hardback) | ISBN 9781119644224 (adobe pdf) | ISBN 9781119644286 (epub)
Subjects: MESH: Biomedical Technology | Biomedical Engineering | Technology Assessment, Biomedical
Classification: LCC R855.3 (print) | LCC R855.3 (ebook) | NLM W 82 | DDC 610.28–dc23
LC record available at https://lccn.loc.gov/2020025470
LC ebook record available at https://lccn.loc.gov/2020025471

Cover Design: Wiley
Cover Image: © CI Photos/Shutterstock

Set in 9.5/12.5pt STIXTwoText by SPi Global, Chennai, India

10 9 8 7 6 5 4 3 2 1

To our parents

Contents

List of Contributors

Rami Ghannam,
University of Glasgow
Glasgow
UK

Muhammad A. Imran
University of Glasgow
Glasgow
UK

Qammer H. Abbasi
University of Glasgow
Glasgow
UK

You Hou
University of Glasgow
Glasgow
UK

Yuchi Liu
University of Glasgow
Glasgow
UK

Yidi Xiao
University of Glasgow
Glasgow
UK

Guodong Zhao
University of Glasgow
Glasgow
UK

Francesco Fioranelli
Technical University of Delft
Netherlands

Julien Le Kernec
University of Glasgow
Glasgow
UK

Ahmed Zoha
University of Glasgow
Glasgow
UK

Janet Bouttell
University of Glasgow
Glasgow
UK

Eleanor Grieve
University of Glasgow
Glasgow
UK

Neil Hawkins
University of Glasgow
Glasgow
UK

Naem Ramzan
University of West of Scotland
UK

Introduction

Along with Medicine and Law, Engineering is one of the oldest professions in the world. While there is debate regarding the exact definition of engineering, the word engineer derives itself from the Latin word ingenium, which means ingenuity Wall [2010]. In fact, engineering involves the application of science and technology to develop new products, tools, services or processes that can benefit society Crawley et al. [2007]. According to Theodore Von Krmn: “Scientists discover the world that exists; engineers create the world that never was” Mackay [1991]. Consequently, engineers are problem solvers and are responsible for creating the healthcare products that we see today.

During the past two hundred and fifty years, the field of engineering has witnessed several waves of innovation, which have depended on the world’s techno-economic paradigm shifts Perez [2010]. Each of the these overlapping waves are approximately 50 years in duration and are also known as the Kwaves. Thanks to the use of iron, waterpower and mechanical constructions, the first of these innovation waves started with the industrial revolution in 18th Century Britain de Graaff and Kolmos [2014]. Now, as we approach the 21st century and the sixth innovation wave, engineers are shifting their interests from the fields of physics, electronics and communications to the interdisciplinary fields of biology and information technology.

Thus, during the past two decades the healthcare industry has seen a rapid transformation. In fact, medical technologies have evolved since the development of the bifocal lens in the early 18th century by Benjamin Franklin. Today, engineers are transforming these contact lenses into healthcare platforms that can monitor vital human signs Yuan et al. [2020]. Consequently, the healthcare field is continuously being reshaped through advances in sensors, robotics, microelectronics, big data and artificial intelligence.

Innovation in healthcare is now a widely researched topic. It is currently a “hot” topic, since it is desperately needed. Without doubt, innovation allows us to think differently, to take risks and to develop ideas that are far better than existing solutions. In this book, we aim to highlight the research that engineers have been engaged in for developing the next generation of healthcare technologies.

Currently, there is no book that covers all topics related to microelectronics, sensors, data, system integration and healthcare technology assessment in one reference. This book aims to critically evaluate current state-of-the-art technologies and will provide readers with insights into developing new solutions.

The book discusses how advances in sensing technology, computer science, communications systems and proteomics/genomics are influencing healthcare technology.

Our book is highly beneficial for healthcare executives, managers, technologists, data scientists, clinicians, engineers and industry professionals to help them identify realistic and cost-effective concepts uniquely tailored to support specific healthcare challenges. Moreover, researchers, professors, doctorate and postgraduate students would also benefit from this book, as it would enable them to identify open issues and classify their research based on existing literature. In fact, these academics need to ensure that their curricula are constantly being revised and updated according to the previously mentioned innovation waves Ahmad et al., Magjarevic et al. [2010], Xeni et al..

Additionally, our book aims to provide in-depth knowledge to stakeholders, regulators, institutional actors, research agencies on the latest developments in this field, which serves as an aid to making the right choices in prioritizing funding resources for the next generation of healthcare technologies. The first chapter deals with Healthcare Technology Assessment (HTA). This chapter focuses on three main topics. The first aims to provide an explanation of the principles of HTA and its familiar role in determining coverage of health care provision. The second involves outlining the challenges of health technology assessment for medical devices. An outline of the main categories of devices will be presented (large capital items, point of care devices, diagnostics, implantables and telehealth) and the difficulties associated with evaluating each of these types of devices. Challenges include licensing and regulation, incremental improvement, evidence generation, short lifespan, workflow, behavioural and other contextual factors and indirect health benefit. Finally, the authors will mention the contribution of HTA in the development and translation of medical devices. They will set out the role of HTA in identifying needs, assessing the potential of technologies in development, aiding design and tailoring evidence generation activities. The chapter will also be Illustrated with appropriate case studies.

Chapter 2 deals with contactless RF sensing, which has recently gained plenty of interest in the domain of healthcare and assisted living due to its capability to monitor several parameters related to the health and well-being of people. This ranges from respiration and heartbeat to gait and mobility, to activity patterns and behaviour. The main advantage of RF sensing is its contactless monitoring capability. Consequently, no sensors need to be worn by the person monitored and no optical images need to be taken via conventional cameras, which can raise problems of privacy especially in private homes. The aim of this chapter is to provide an overview of the most recent different RF technologies for healthcare, including active and passive radar and wireless channel information.

Chapter 3 discusses recent advances in Pervasive Sensing. Here, the vision of nanoscale networking attempts to achieve the functionality and performance of the Internet with the exceptions that node size is measured in cubic nanometres and channels are physically separated by up to hundreds or thousands of nanometres. In addition, these nano-nodes are assumed to be self-powered, mobile and rapidly deployable in and around a specific target. Nevertheless, downscaling the principles of traditional electromagnetic networks to the nanoscale introduces several challenges, both in terms of device technologies and communication solutions. This chapter will shed light on the basic principles of nano-electromagnetic communication in the Terahertz frequency region in the nanoscale dimension.

Moreover, chapter 4 is concerned with providing recent advances in Microelectronics for Brain Implants. This Chapter discusses advances in diagnosis, monitoring, management and treatment of neurological disorders. It will be two parts: first we will discuss our approaches for in vitro diagnostics include lab-on-chip progresses for neurodegenerative diseases such as Alzheimers and Parkinsons diseases. Secondly, we will review our in-vivo implantable medical devices for different applications include treatments of epilepsy and spinal cord. We will conclude this chapter from different perspective including sensing, communications and energy harvesting.

Chapter 5 describes the rationale for using machine learning (ML) techniques for decision making in the healthcare industry. In human physiology, hydration is essential for the proper functioning of multiple systems. Hydration is responsible for controlling various biological reactions by acting as a solvent, a reaction medium, a reactant and a reaction product. Water is the major component of the human body, making it critical for thermoregulation, cell volumes and even for joint lubrication. This chapter will deal with applying machine learning techniques on data collected from a controlled environment for detection of skin hydration levels.

In chapter 6, the authors describe how machine learning techniques can revolutionize medical diagnosis. Single Nucleotide Polymorphisms (SNPs) are one of the most important sources of human genome variability and ML has the potential to predict SNPs, which can enable the diagnosis and prognosis of several human diseases. To separate the affected samples from the normal ones, various techniques have been applied on SNPs. Achieving high classification accuracy in such a high-dimensional space is crucial for successful diagnosis and treatment. In this work, we propose an accurate hybrid feature selection method for detecting the most informative SNPs and selecting an optimal SNP subset.

Chapter 7 provides an overview of the energy harvesting techniques that can be used to power wearable and portable devices. Power harvesting or generation is still a big challenge in biomedical devices. Conventionally, these devices have relied on batteries, which are disadvantageous due to their size, lifespan and hazardous nature. It is therefore important to investigate alternative methods to ensure that medical devices are battery-free and/or are self-powered. There are various methods that are currently being investigated, which include Photovoltaic cells (PV), Piezoelectric Generators (PEG), Thermoelectric Cells (TEG) and Radio Frequency techniques. This chapter aims to provide the latest trends in each of these energy harvesting methods and offer recommendations for future applications.

Chapter 8 is concerned with how data can be transferred from in-vivo to in-vitro Due to losses in human tissue, reliable data transfer through skin is a major challenge. In this chapter, a novel concept of cooperative communication for in-vivo nano network will be presented to enhance the communication among these devices. The effect on the system outage probability performance will be conducted for various parameters including relay placement, number of relays, transmit power, bandwidth and carrier frequency.

Chapter 9 aims to provide a rationale for Wireless Control. In this chapter, we will discuss the real-time wireless control of life-critical actions, which is one of the essential features to enable many healthcare applications, e.g., remote diagnosis and surgery. In particular, we will introduce the basics of wireless control systems and discuss the fundamental design capabilities needed to realize real-time wireless control, with primary emphasis given to communicationcontrol co-design. The goal is to provide integrated solutions for life-critical actions in healthcare.

Similarly, chapter 10 deals with how life-critical communications networks can be developed. It is noteworthy to mention that device-to-device (D2D) communication is regarded as a promising solution to improve the spectrum utilization of cellular systems. This is due to the direct link between nearby devices, which can be established on the same time/frequency resources (cellular resources) over a short distance. Consequently, D2D communication can be a potential candidate for mobile (M)-health applications. However, due to their intrinsically open nature, D2D communication is vulnerable to security attacks. In this chapter, we will highlight the security requirements of D2D communication to make them applicable to M-health scenarios. In addition, we will investigate the standardization efforts for secure D2D communications.

Finally, concluding remarks are provided in the final chapter of the book, where we discuss the challenges and opportunities ahead for healthcare devices and systems.

References

Wasim Ahmad, Rami Ghannam, and Muhammad Imran. Course design for achieving the graduate attributes of the 21st century uk engineer. In Poster presented at Advance HE STEM Teaching and Learning Conference, Birmingham, UK. 30-31 Jan 2019.

Edward Crawley, Johan Malmqvist, Soren Ostlund, Doris Brodeur, and Kristina Edstrom. Rethinking engineering education. The CDIO Approach, 302:60–62, 2007.

Erik de Graaff and Anette Kolmos. Innovation and research on engineering education. In Handbook of research on educational communications and technology, pages 565–571. Springer, 2014.

Alan L Mackay. A dictionary of scientific quotations. CRC Press, 1991.

Ratko Magjarevic, Igor Lackovic, Zhivko Bliznakov, and Nicolas Pallikarakis. Challenges of the biomedical engineering education in europe. In 2010 annual international conference of the IEEE engineering in medicine and biology, pages 2959–2962. IEEE, 2010.

Carlota Perez. Technological revolutions and techno-economic paradigms. Cambridge journal of economics, 34(1):185–202, 2010.

Kevin Wall. Engineering: Issues, challenges and opportunities for development. UNESCO, 2010.

Nikolas Xeni, Rami Ghannam, Fikru Udama, Vihar Georgiev, and Asen Asenov. Semiconductor device visualization using tcad software: Case study for biomedical applications. In Poster presented at IEEE UKCAS 2019, London, UK. 6 Dec 2019.

Mengyao Yuan, Rupam Das, Rami Ghannam, Yinhao Wang, Julien Reboud, Roland Fromme, Farshad Moradi, and Hadi Heidari. Electronic contact lens: A platform for wireless health monitoring applications. *Advanced Intelligent Systems*, page 1900190, 2020.

1

Maximizing the Value of Engineering and Technology Research in Healthcare: Development-Focused Health Technology Assessment

Janet Boutell Hawkins and Eleanor Grieve

Institute of Health & Wellbeing, University of Glasgow, Glasgow, UK

This chapter focuses on three main topics. The first aims to provide an explanation of the principles of health technology assessment (HTA) and its familiar role in determining coverage of healthcare provision. Second, we discuss the growing contribution of HTA in the development and translation of medical devices introducing what we term "development-focused HTA"(DF-HTA). We set out the role of DF-HTA in identifying needs, assessing the potential of technologies in development, aiding design, and tailoring evidence generation activities. Finally, we outline the challenges of development and assessment presented by medical devices distinguishing large capital items, point of care devices, diagnostics, implantables, and digital devices. Each category of device has its own set of challenges for developers and HTA analysts alike. Challenges include a complex licensing and regulation environment, short lifespan and incremental improvement, difficulties in generating clinical evidence, the importance of contextual factors (e.g., how the device will be used and by whom), patient and clinician acceptance, and the indirect health benefit from diagnostic devices.

1.1 Introduction

Advancements in engineering and technology have the potential to revolutionise patient care and medical research. However, resources available for research and development and for healthcare provision are limited, so it is essential that any funds invested are spent on those projects that are both likely to succeed and likely to make a difference to patients' health. Health Technology Assessment (HTA) is a multi-disciplinary approach that studies the medical, social, ethical, and economic implications of development, diffusion, and use of health technology (INAHTA.ORG 2019). HTA has been most widely used by public payers or reimbursement agencies when a technology (such as a pharmaceutical or a medical device) is ready for market. However, there is increasing recognition that HTA undertaken at an earlier stage in the development of a health technology can aid investors and developers to focus their resources on technologies that are likely to succeed as well as identifying those that are likely to fail (IJzerman et al. 2017). We term this earlier form of HTA,

Engineering and Technology for Healthcare, First Edition.
Edited by Muhammad Ali Imran, Rami Ghannam and Qammer H. Abbasi.

"development-focused HTA" (DF-HTA) and the more familiar form of HTA "use-focused HTA."

Health technology is a broad term that encompasses drugs, medical procedures, tests, and service configuration. Medical devices form a sub-set of health technology. The diverse sub-set includes large, expensive, capital equipment such as the Da Vinci robotic surgery platform (INTUITIVE.COM 2019) and small consumable items such as sticking plasters. There are some common challenges for developers of all categories of medical device. In particular, the licensing and regulatory environment is highly complex and differs according to the jurisdiction where the device will be used. Evidence generation is also particularly challenging for many kinds of medical devices as different decision-makers require different levels of evidence. For devices with short lifespans, when it is common for different versions to be developed sequentially with incremental improvements, it is difficult to know which version of the device the evidence relates to. Items like the robotic surgery platform are subject to the "learning curve" effect, as surgeons need an initial training period to improve their competence before the clinical effectiveness of the new equipment can reasonably be compared with prior standards of care. Diagnostic tests form an important sub-category of medical devices. Evidence generation for diagnostics is challenging because any health outcome resulting from the use of the diagnostic is indirect rather than direct. In order for there to be an improvement in health, the diagnostic test needs to change the diagnostic or treatment pathway so that the patient is treated sooner or more effectively. Not only is any health gain indirect, it also depends upon the behavior of the clinician and the patient. A test may indicate that treatment B is more appropriate for the patient, but if the patient and/or the clinician prefer treatment A, the test cost has been wasted and the patient's health is not improved. The value proposition for many devices is also contextually dependent. By this we mean that the device may add value in some places but not others, depending on factors such as what the current treatment and diagnostic pathways are; staffing levels; capacity and workflow; and, what other capital equipment is in place.

The numerous challenges facing developers of medical technologies in general, and medical devices in particular, have led to a recognized problem in translating research from bench to bedside. One response to this has been the growth of translational research bodies charged with supporting developers and bridging the translation gap. Two notable contributors to the DF-HTA literature are the Center for Translational Molecular Medicine (LYGATURE.ORG 2019), based in the Netherlands and MATCH UK (MATCH.AC.UK 2018), a collaboration between several UK universities. This growing literature demonstrates how the various challenges of medical device development can begin to be addressed at an early stage of development using the methods of DF-HTA.

The aims of this chapter are to explain what HTA is and how it has been used to determine the coverage of healthcare provision; to explain what DF-HTA is and how it differs from use-focused HTA; to set out the challenges in the development and assessment of medical devices; and to illustrate the contributions of DF-HTA in the development and translation of medical devices through a number of case studies.

1.2 What Is HTA?

Healthcare resources are limited in every setting, and decision-makers are faced with difficult choices about which technologies should be adopted and used within their service. The definition of HTA given in the introduction (INAHTA.ORG 2019) was

> HTA is a multi-disciplinary approach which studies the medical, social, ethical and economic implications of development, diffusion and use of health technology.

Technology in HTA is widely defined and includes drugs, devices, health services, and systems. As the study of these various aspects of health technologies, HTA is well-placed to inform decision-makers as they make resource allocation decisions. Indeed, the role of HTA to inform decision-makers is included in the World Health Organisation (WHO.INT 2019) definition of HTA:

> the systematic evaluation of properties, effects and/or impacts of health technologies and interventions. It covers both the direct, intended consequences of technologies and interventions and their indirect, unintended consequences. The approach is used to inform policy and decision-making in health care, especially on how best to allocate limited funds to health interventions and technologies.

An ongoing project to reach a consensus definition of HTA proposed a definition that includes the important additional factors of a systematic and transparent process.

> a multidisciplinary process that uses explicit and scientifically robust methods to assess the value of using a health technology at different points in its lifecycle. The process is comparative, systematic, transparent and involves multiple stakeholders. The purpose is to inform health policy and decision-making to promote an efficient, sustainable, equitable and high-quality health system.

Health Technology Assessment, as a discipline, first developed in the United States when Congress requested Technology Assessment of health technologies in the mid 1970s (Stevens et al. 2003), and the term is now internationally used. The adoption of this term gained popularity in wealthier countries that prioritized the evaluation and improvement of health care. HTA draws on Evidence Based Medicine (EBM). EBM developed from the publication in 1972 of Archie Cochrane's "Effectiveness and Efficiency" (Cochrane 1972) and is now championed by the international organization, the Cochrane Collaboration (Stevens et al. 2003). Evidence synthesis methods such as systematic review and meta-analysis are core to HTA and draw heavily on guidance developed by the Cochrane Collaboration. These methods often form the basis for the clinical effectiveness estimates in cost-effectiveness analysis and health economic modelling.

The components of HTA vary according to the particular decision-maker, but many forms of HTA start with the definition of a decision problem to address. Analysts may find it useful to use a structure to help them define the decision problem. A popular structure is PICO, which stands for Population, Intervention, Comparator, and Outcome. The intervention is the technology to be assessed, and the comparator is the current standard of care in that disease area. Once the decision problem has been defined, the next step is synthesis of the clinical evidence, using techniques such as systematic review and meta-analysis. Once the evidence on clinical effectiveness has been assembled and issues regarding evidence quality and generalisability addressed, cost-effectiveness can be considered. Finally, other considerations such as legality and ethics may be addressed (Eddy 2009).

HTA informs a variety of healthcare decision-makers, ranging from national healthcare providers like the National Health Service in the UK, to regional health authorities (for example, in Spain and Canada) and local providers such as hospitals. Insurance companies and commercial healthcare providers also need to make decisions about coverage and reimbursement. HTA agencies may be established within, or supported by, the decision-maker as with the National Institute for Health and Care Excellence (NICE) in the UK or may be external bodies such as the Institute for Clinical and Economic Review (ICER) in the United States, which is funded primarily by not for profit organizations (ICER.ORG 2019) and provides advice for guidance. Some agencies have a strong emphasis on cost-utility analysis (for example, UK, Netherlands, Canada) and some have acknowledged a financial limit to the amount they consider acceptable to pay for each year in full health delivered by a health technology.

Decisions supported by HTA include two broad categories: allocation of a set budget over a number of healthcare areas and decisions about individual technologies or programs. In the first category, the decisions involve which programs to include in a package of Universal Health Coverage (for example, maternity care, vaccination programmes) or decisions about prioritization within a research budget. The aim of the HTA would be to allocate the budget according to agreed criteria of effectiveness, value for money, and other considerations, perhaps equity. The second category includes assessment of individual technologies, such as pharmaceuticals, to determine whether they should be adopted. Again, they would be likely to be assessed against pre-established criteria relating to evidence base, need, value for money, and equity issues. Medical devices and surgical procedures could also be assessed in this way. HTA may also be used to determine whether a technology in current use should be excluded from reimbursement or coverage. This is known as "disinvestment." There is growing interest in how HTA could be used before or during the development of a technology to inform a broader set of decisions. This form of HTA, which we have termed "development-focused HTA," is the subject of the next section.

1.3 What Is Development-Focused HTA?

Development-Focused HTA (DF-HTA) is concerned with whether and how a technology should be developed. It is contrasted in this section with Use-focused HTA (described in the previous section), which compares the clinical benefit that an available technology is likely to deliver to the cost of the technology and makes a recommendation to a decision-maker based on an assessment of opportunity cost and other local criteria. DF-HTA differs in that the technology is under development, perhaps still at the concept stage. DF-HTA aims to inform the developers of the technology about a wider range of questions, including how the technology should be designed, used, and/or priced. DF-HTA is a relatively young but expanding field. We believe that the tools of DF-HTA could be usefully employed to evaluate technologies in development and ensure that only those that are likely to succeed continue

Table 1.1 Key Differences between Use-Focused and Development-Focused HTA.

Feature	Use-focused HTA	Development-focused HTA
Target audience	Reimbursement agencies, insurers, clinicians & patients	Technology developers, investors, and public sector funders
Specific decisions HTA designed to inform	One off Binary - accept/reject Optimising guidelines Price revisions Reimbursement decisions Budget allocation	Broad range including: Go/no-go decisions Technology design Trial design Research prioritization Reimbursement strategy prioritization
Available evidence	Evidence-base more developed	Evidence-base fluid, may be limited
Timing	Close to and post-approval	Repeated Pre- and during development
Underlying user objective	Maximize health	Maximize financial and/or societal return on investment
Core decision rule	Reimburse when value meets established criteria	Continue development if project has (most) potential to deliver financial/societal return on investment
Clinical decision space	Targeted at specific decision-makers, indications, comparator and patient groups defined by local practice and licensing	Potentially multiple jurisdictions, indications, comparators, funders, user, groups, thresholds (test cut-off), levels of test performance, and positions in pathway
Business model	Fixed, reimbursement by payer/insurer	Broad, not yet defined
Resources for analysis	Committed - limited number of technologies reviewed	Often constrained
Stance of analysis	Normative	Positive
Burden of Proof	Established standard of evidence	Evidence credible to the development team

to be developed as well as to prioritize research expenditure on new health technologies. This form of HTA, used to inform developers of health technologies, has been termed "early HTA" in the academic literature (Ijzerman and Steuten 2011). We prefer to use the label "development-focused HTA" as it is the audience, rather than the timing of the HTA, which drives many of the differences. Table 1.1 sets out key differences between use-focused and development-focused HTA, and these are explained in the paragraphs that follow using an example of home brain monitoring in epilepsy patients.

1.4 Illustration of Features of Development-Focused HTA

Digital health technologies have the potential to improve patient health through improved diagnosis and/or ongoing monitoring of health conditions. They may also reduce cost by accelerating diagnosis and/or reducing hospital admissions. The following paragraphs contrast a use-focused HTA exercise and a development-focused exercise concerning a home-based brain monitoring device (HBM) for epilepsy patients. Epilepsy is diagnosed using electroencephalography (EEG), but because the standard routine of EEG is relatively short, it has only 20-56% sensitivity (Breteler 2012). HBM could increase the sensitivity of diagnosis to in excess of 90 percent by increasing the period of observation and adding a detection algorithm (Breteler 2012).

1.4.1 Use-Focused HTA

The timing of any use-focused HTA would be after the device was licensed when it was available for purchase. The audience for a use-focused HTA of HBM would generally be a national decision-maker (perhaps a ministry of health or a national reimbursement agency) but could potentially be a healthcare provider at a more local level, such as a hospital. The underlying objective of either of these decision-makers would be to maximize health given the budget at their disposal. The specific decisions to be informed would be whether to purchase the monitoring system and potentially, in which populations it should be used. Although the level of evidence required for a device to be licensed is not as well-defined as the evidence required for a pharmaceutical, the available evidence should include evidence of safety and performance. The analysts undertaking the use-focused HTA may find sufficient evidence of clinical utility or may flag up that some additional evidence is required prior to any decision being made. The price of the system would be known (although it may potentially be open to negotiation). The decision space in this exercise would be relatively narrow as the jurisdiction and disease are both fixed. It would probably be necessary to consider different sub-groups of the population where the effectiveness of HBM may vary and possibly different positions in the diagnostic pathway. A use-focused HTA of a diagnostic technology may involve modelling the impact of the technology (HBM) on health outcomes and resource use over a long time-horizon. This analysis would produce an estimate of the clinical efficacy and cost-effectiveness of the technology, which could then be used to inform the one-off decision about whether or not to purchase. This decision would be informed by the decision-makers' underlying decision rule in order to decide whether or not to implement the program. If a national decision-making body commissioned the HTA exercise, the resources available for analysis are likely to have been adequate to undertake a comprehensive analysis. However, for diagnostic technologies and other devices, use-focused HTA is sometimes undertaken by smaller, local healthcare services, and they may need to undertake a less comprehensive review to reflect the resources they have available.

1.4.2 Development-Focused HTA

By way of contrast, the timing of a development-focused HTA may precede the discovery research for a digital health application or may be undertaken when there is a prototype available but there are decisions to be made about whether it is worthwhile to continue the development or to prioritize its implementation. The target audience for the HTA analysis may be a public or charitable research funder allocating funds across a portfolio of projects or a commercial developer/investor. The underlying objective of these two groups would potentially differ with public or charitable research funders looking to maximize health given the budget at their disposal and commercial developers/investors looking to maximize financial return on investment. The timing of the assessment would determine the available evidence, but this is unlikely to be large-scale technology-specific evidence even at the prototype stage. Evidence is more likely to come from bench studies, similar technologies, or assumptions informed by input from experts. In contrast to use-focused HTA, the decision space in this exercise may be very wide. There may be scope to use HBM in many different geographical areas, populations, and diseases. As in use-focused HTA, the analysis may involve modelling the health and cost impact of HBM, but a number of plausible scenarios may be modelled incorporating evidence and assumptions as described above. Modelling would be an iterative process, revised a number of times, reflecting evidence generated and with increasing sophistication as the decision-space became narrower through the development process. Rather than informing a single decision, the analysis informs ongoing discussions. Even if the analysis showed that the technology in the current form in the selected scenarios did not look promising, this may not mean that it should be abandoned - it may instead indicate that other settings are preferable or an improved design is required.

1.5 Activities of Development-Focused HTA

The features identified above drive important differences between development and use-focused HTA in the analytic tools used. Synthesis of clinical evidence and economic evaluation are the mainstay of use-focused HTA, as the assessment tends to focus on comparative effectiveness, cost-effectiveness, and budget impact associated with well-defined interventions. In contrast, development-focused HTA draws on a broad range of multi-disciplinary methods due to the wide range of the decisions that development-focused HTA is intended to inform. Development-focused HTA can be considered as contributing to iterative and interlinked assessments of clinical and economic value. The clinical value assessment considers what impact the technology might have on clinical practice and health (and wider social) outcomes. The economic value assessment builds on the clinical value assessment to consider the economic impact of changes in healthcare resource use and other economic value drivers such as productivity effects. It may also include consideration of the potential pricing of the new technology, volume of sales, fixed and variable costs of production and distribution in order to produce estimates of net margin. These assessments are informed by and in turn provide information to the research and development process and the business case for the technology (see Figure 1.1).

The arrows in Figure 1.1 indicate the complexity of the links between the aspects of the framework. Table 1.2 sets out examples of the links represented by the arrows, many of which are reciprocal. Rather than being bidirectional, clinical value assessment must necessarily precede economic value assessment as the change in clinical pathways forms the basis of any decision model. The relationship between these two assessments is, however, iterative as economic value assessment (informed by strategic and market analysis undertaken in business case development) may inform the selection of jurisdictions to be assessed in a clinical value assessment.

The relationship between the research and development process and clinical case assessment is reciprocal. Evidence generated in the research and development process is used to

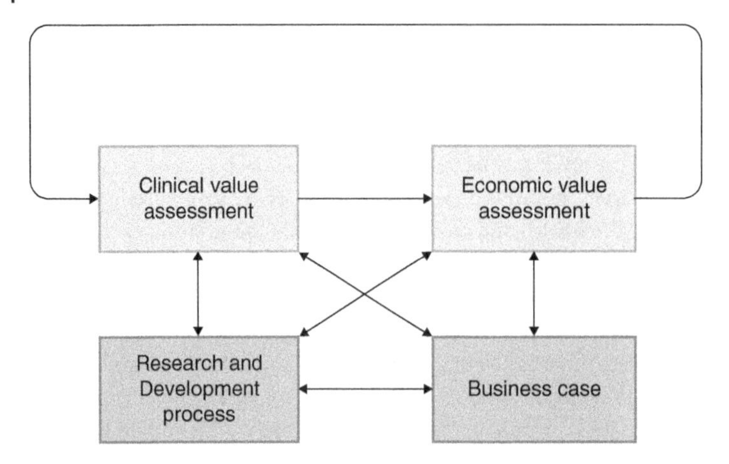

Figure 1.1 Activities of development-focused HTA.

populate the clinical value assessment. Where this evidence is not available (as in much early stage development-focused HTA), methods of evidence generation can be used to fill the gaps (see Table 1.4). It is important to distinguish research and development evidence generation (such as clinical trials and clinical trial simulation) and evidence generation to support the assessment process (such as expert elicitation and the use of estimates from the literature). The clinical value assessment informs the research and development process about the kind of evidence that is required for the assessment as well as providing insight into contextual aspects that may require building into the design and indicating target thresholds of performance for the technology to add clinical value. Business case development may feed target markets and indications directly into the clinical value assessment. The detailed epidemiological analysis involved in clinical value assessment may lead to better-informed estimates of the market size in the business case development.

Economic value assessment may inform the research and development process by using threshold analysis to determine the minimum performance characteristics for a technology to be clinically effective and cost-effective in a particular jurisdiction. Again, this relationship is reciprocal as the research and development process may inform the economic value assessment about the likely cost-profile of the technology. Economic value assessment also has a reciprocal relationship with business case development. Economic value assessment can contribute to pricing decisions as well as providing analysis as to the scale of potential diffusion-related issues. As the business case develops, this can narrow down the indications and jurisdictions to be targeted and forming the basis for the economic value assessment. The final reciprocal relationship shown in the proposed framework is that between the research and development process and business case development. The research and development process informs business case development about the potential indications that the technology could target as well as information about the cost profile of the technology and development costs. Business case development provides guidance to the research and development process about the thresholds for these cost parameters for the technology to be commercially viable.

Table 1.2 Examples of information flows represented by arrows in Figure 1.1

Arrow from	Arrow to	Information flow
Clinical value assessment	Economic value assessment	Clinical effectiveness of technology and comparator – multiple jurisdictions/indications, Contextual information to allow modelling of pathway changes, Insight into diffusion issues that may be included in modelling – e.g., importance of adherence
Economic value assessment	Clinical value assessment	Markets/indications with most economic potential
Clinical value assessment	Research and development process	Evidence gaps Threshold technology performance to improve clinical effectiveness Insights into contextual aspects requiring consideration in design
Research and development process	Clinical value assessment	Evidence of clinical effectiveness
Clinical value assessment	Business case development	Market size
Business case development	Clinical value assessment	Target markets and indications
Economic value assessment	Research and development process	Threshold technology performance for cost-effectiveness Evidence gaps
Research and development process	Economic value assessment	Cost profile of technology
Economic value assessment	Business case development	Pricing thresholds Insight on potential impact of diffusion-related parameters
Business case development	Economic value assessment	Target markets and indications for scenario selection
Research and development process	Business case development	Cost profile of technology – development and production Potential indications
Business case development	Research and development process	Thresholds for costs of production and development

1.6 Analytical Methods of Development-Focused HTA

Table 1.3 presents analytic methods of development-focused HTA that have been employed in undertaking the clinical and economic value assessments, as well as methods that have been previously classed as DF-HTA but which we would classify as either research and development or business case development. Table 1.4 presents methods of evidence generation for DF-HTA. We illustrate a selection of appropriate methods through the continuation of the HBM example and through the case studies that will be discussed later in the chapter.

Table 1.3 Analytical methods by framework aspect

Framework aspect	Analytical methods
Clinical value assessment	Epidemiological analysis Health impact assessment
Economic value assessment	Cost-effectiveness analysis (including cost-utility analysis, cost-consequence analysis, cost-benefit analysis, cost-minimization analysis, headroom analysis, probabilistic sensitivity analysis, and one-way sensitivity analysis) Value of Information analysis (including EVPI, EVPPI, EVSI)
	Multi-criteria decision analysis (Analytic Hierarchy Process)
Research and development process	Clinical trials Clinical trial simulation Bench studies User-centred design
	Technological forecasting based on epidemiological data
	Research and development portfolio management Failure and reliability analysis
	Technology profiling/uncertainty profile
	Brainstorming sessions
	Preliminary market research
Business case development	Strategic planning methods: PEST, SWOT
	Horizon-scanning
	Scenarios building and evidence profile
	Return on investment Soft systems methodology
	Payback from research analysis
	Real Options Analysis
EVPI	Expected Value of Perfect Information
EVPPI	Expected Value of Partial Parameter Information
EVSI	Expected Value of Sample Information
PEST	Political, Economic, Social, Technological
SWOT	Strengths, Weaknesses, Opportunities, Threats

Table 1.4 Evidence generation methods to support development-focused HTA

Method	Examples
Literature review/analysis	Archives, documents
Qualitative methods of user interaction	Focus groups, workshops, questionnaires, interviews, and surveys
Choice-based preference methods	Discrete choice experiments and conjoint analysis
Multi-criteria decision analysis	Analytic Hierarchy Process
Expert opinion and structured expert elicitation	Delphi method

1.6.1 Clinical Value Assessment

The clinical value assessment essentially compares clinical practice and outcomes in two or more future worlds: one without the new technology and one or more with (if there are multiple options for employing the new technology). The assessment of clinical value requires an understanding of the epidemiology of the disease, its current treatment and expected costs, and outcomes in the local context. This may be based on published epidemiological studies and published treatment guidelines, opinions from local experts, and, if feasible, local primary research. In all cases, it is crucial that the evidence be relevant to the setting(s) that we are interested in. In practice, this epidemiological analysis in relation to HBM would involve understanding the prevalence of relevant diseases, sub-types, and populations affected in a variety of geographical locations. Information required would be the number of patients affected, the impact on their health, and the current diagnostic and treatment pathways. These may be many and various depending upon the context. Practical considerations may require a narrowing of the settings investigated in-depth, although it is useful to have an overview of all potential relevant areas.

Next, a health impact assessment of the potential impact of the technology on the disease and treatment pathway will be required. In development-focused HTA, where there is likely to be a paucity of trial data or other empirical data relevant to the new technology, alternative methods of evidence generation are often required. These are described in the Evidence Generation section below. For HBM, we would need estimates of the technology's clinical performance in different positions in the diagnostic pathway (e.g., its ability to correctly identify epilepsy). We would also need to consider how the information from the test changed the diagnostic and treatment pathways and what impact this had on overall health outcomes (i.e., the clinical utility of the technology). An initial health impact assessment may make a range of estimates of clinical performance, including a perfect technology with 100% epilepsy patients identified. This assists developers in setting the boundaries for performance and/or price required for the technology to be clinically efficacious and/or cost-effective.

It is important also to note that new technology may offer incremental value in many ways, each of which may be valued in a different way by different users or other stakeholders. A technology may directly improve health outcomes, may facilitate service, and process improvements and/or reduce resource use. There may, for example, be value in a technology that is not as effective as the current standard of care but can be delivered in areas where the power supply is inconsistent. Ease of operation is another aspect of value that may need to be taken into account. Technologies may add value through being light and portable, using readily available consumables or having no requirement for the user to be literate.

1.6.2 Economic Value Assessment

The economic value assessment extends the clinical value assessment by accounting for healthcare and other resource usages. It may also include productivity and other wider impacts. In use-focused HTA, economic evaluation is typically undertaken to determine whether the technology, at a given price and given current evidence, represents a cost-effective use of healthcare resources. In development-focused HTA, economic evaluation plays a different role. It is used to support a range of decisions including whether further development should be funded and how studies should be designed.

Cost-benefit, cost-utility, cost-effectiveness, cost-consequence, and cost-minimisation analysis (Drummond et al. 2015), are all forms of economic value assessment in that all compare the difference in outcomes brought by new technology to the difference in costs. All methods calculate costs in the same way; however, they measure outcomes differently. In a cost-benefit analysis, outcomes are measured in monetary terms. The principal challenge is determining how the outcomes should be valued. In a cost-utility analysis, outcomes are expressed in quality-adjusted life years (QALYs) or disability-adjusted life years (DALYs). This form of economic evaluation is preferred in the Gates Reference Case for Low and Middle Income Countries (LMIC) (Claxton et al. 2014). This is because the outcome measures, QALYs and DALYs, used in cost-utility analysis are generic measures that facilitate comparisons across technologies and disease areas. In cost-effectiveness analysis, outcomes are expressed as a single disease-specific measure: for example, cost per infection avoided. In cost-consequence analysis, outcomes are expressed across multiple measures. A particular challenge with both these forms of analysis is the difficulty in making comparisons between indications and determining an acceptable threshold for willingness to pay. Finally, in cost-minimization analysis, it is assumed that outcomes are either equal or superior with the new technology, and hence it is sufficient to simply compare costs. The challenge with this analysis is that it is only appropriate when it is safe to assume that outcomes are either equal or superior with the new technology.

Cost-effectiveness analysis may be used in development-focused HTA in two main ways. It can be used to indicate whether a technology is likely to be regarded as cost-effective, and hence used at a given price for the new technology. Alternatively, it can be used to estimate the maximum price at which the new technology is likely to be deemed cost-effective. This has been referred to as "headroom" analysis (Chapman 2013). Where these analyses are based on cost-effectiveness or cost-utility analysis, an estimate of an acceptable threshold willingness to pay will be required. A threshold of three times a country's per capita Gross Domestic Product (GDP) has historically been used in LMICs. However, research is ongoing to determine values for the acceptable threshold that better reflects the opportunity cost of "shadow price" of investment in other healthcare technologies. Threshold analysis is a similar formulaic approach, which assumes a price for the technology and then investigates what values the other parameters need to take in order for the technology to remain cost-effective. These approaches are very simple and quick to perform and can be based on expert opinions or assumptions so they are ideal for undertaking extensive scenario or sensitivity analysis. Any factors that are not modelled can be assessed qualitatively alongside the simple modelling (Chapman 2013). For example, in the assessment of a diagnostic test for hepatitis C in development, (Chapman 2013) discussed the potential impact of a move away from routine testing by biopsy, which could limit the market for the new test. Scenario analysis examining different settings (for example, high/low prevalence and urban/rural setting) and varying sensitivity and specificity in local settings can be undertaken.

In addition to providing information about the potential economic value of a new technology, cost-effectiveness analyses can be used to provide estimates of the potential value of additional information from further research. This is known as Value of Information (VOI) analysis. The analysis can be used to estimate the net impact on health benefit arising from an increase in the probability of selecting the optimal treatment resulting from further research reducing uncertainty. VOI can aid study design and investment decisions (Vallejo-Torres et al. 2011).

In our example of a development-focused HTA of HBM, a number of scenarios would be worth considering. If commercial developers were developing HBM, they may use a form of headroom analysis based on cost-utility analysis to estimate the price they may be able to charge for the technology in order for it to remain below an acceptable cost-effectiveness threshold. They could use threshold analysis to determine what clinical performance would be needed to justify a given price. They could use value of information analysis to design clinical trials which address the areas of most uncertainty (Vallejo-Torres et al. 2011; Wong et al. 2012; Meltzer et al. 2011). A cost-utility analysis undertaken to inform a charitable or public sector funder may assume a price sufficient to cover the ongoing unit cost of the technology and then calculate the incremental cost-effectiveness ratio of different positions in the diagnostic pathway or different diseases. If the development of the technology seemed likely to have an acceptable incremental cost-effectiveness ratio (i.e., to deliver sufficient health impact for its cost), then the project should be continued. If it is too expensive, developers may need to try to alter the design to improve effectiveness or acceptability to patients and clinicians. VOI analysis has been used in a number of research prioritization pilots to help funders choose which clinical trials to fund (Bennette et al. 2016; Carlson et al. 2013; Meltzer et al. 2011). Alternatively, the results of the cost-utility analysis may form part of a multi-criteria analysis for project prioritization or the basis for a calculation of societal return on investment.

The economic analysis can be broadened to consider financial return on investment (ROI) and/or social return on Investment (SROI). ROI is used by commercial entities to estimate the likely return on a project and to determine whether to continue the project or not. The basic calculation is: expected revenues; less ongoing costs of production/distribution; less development costs still to be incurred. The commercial viability of a candidate technology will depend on combinations of price and volume that are achievable, fixed and variable costs of production, and development costs (Meltzer et al. 2011). The price or cost estimates from the headroom calculation can be multiplied by the expected volume of sales (from epidemiological analysis) to calculate revenue. SROI is a framework for measuring and accounting for a broader concept of value; it considers aspects valued by different stakeholders, such as reductions in inequality, environmental degradation, and improvements in well-being by incorporating social, environmental and economic costs and benefits (SOCIALVALUEUK.ORG 2019). Gains are typically measured in monetary terms (financial ROI) or can also be expressed in terms of social values. Costs remain the same in both cases. A societal return on investment analysis can be undertaken from multiple stakeholder perspectives (patient, society, manufacturer). Both ROI and SROI can be used as a means to prioritize projects within portfolios generally as input to a multi-criteria process. For example, a research team looking at investments in the development of biomarkers in the prevention of type 2 diabetes considered attributes including reduction in downstream costs, added quality-adjusted survival, cost of test, feasibility of treat-all option, competition and ease of implementation in an MCDA exercise (de Graaf et al. 2015).

Scenarios-building involves the development and consideration of multiple scenarios reflecting uncertainties about the context and future events. These scenarios could inform stakeholder analysis and be built in to future economic evaluation in order to provide a range of options for consideration throughout the development process (Joosten et al. 2016). This is a good illustration of development-focused HTA's role in informing discussions and decisions throughout the development process.

1.6.3 Evidence Generation

The assessments involved in development-focused HTA may involve consultation with a wide range of stakeholders including patients, physicians, and policy-makers. In the very early stages, evidence may be primarily based on literature review and consultation with the technology development team. Consultations with stakeholders may involve informal or formal qualitative methods such as focus groups or interviews, or may take the form of structured expert elicitation. Informal methods are appropriate (and may be the only option) when resources are limited. A particular concern in expert elicitation is that estimates are often too optimistic (particularly if the expert is involved with the development team) and under-represent uncertainty. Thus, formal methods have developed that attempt to address this concern. Formal methods can be divided into two main groups: structured expert elicitation (SEE) and multiple criteria decision analysis (MCDA). SEE is used to obtain estimates of relevant population parameters, for instance, what is the expected treatment effect for a new technology. MCDA provides a solution to a decision-problem, for instance, which is the optimum choice among a set of treatment options with different characteristics. Delphi methods are commonly used to elicit expert opinions and typically involve two rounds of questions to individual experts (Gosling 2014). In the first round, experts respond without knowledge of other responses. Responses are then shared and a second round of estimates is sought from the entire group. Estimates may include probability distributions or ranges. It is possible to compensate for optimism-bias and under-estimation of uncertainty through sensitivity analysis in economic evaluation or qualitative consideration of a broader range of options in scenario analysis. Responses may also be weighted according to experience in the field. Best practices for the conduct of expert elicitation are discussed by (Iglesias et al. 2016). Several on-line tools are available including SHELF (Gosling 2018) and MATCH uncertainty elicitation (Morris et al. 2014). The availability of these tools will aid the conduct of expert elicitation if resources are limited. Structured methods of expert elicitation have been used to: identify current treatment pathways and standard of care (Davey et al. 2011); to specify the technical features of current technology (Terjesen et al. 2017); to estimate likely levels of test performance (Haakma et al. 2011); to estimate treatment effect (Girling et al. 2007; Ostergaard and Mldrup 2010); to estimate the effect of a new test on discharge rates (Kip et al. 2018); and to consider the likely position of a test in a diagnostic pathway (Breteler 2012).

Multi-Criteria Decision Analysis (MCDA) describes approaches that seek to take explicit account of multiple criteria in a decision-making process (Wahlster et al. 2015). In the context of development-focused HTA, experts will typically be asked to indicate their preferences between current and one or more versions of the new technology. A set of criteria relevant to the decision problem will be identified. Each option will be scored on each criterion. Then, in consultation with the experts, weights will be assigned to each criterion and some form of aggregation conducted to identify the optimum treatment. MCDA techniques have been used to support development-focused HTA in various ways. Several studies have used the techniques to support design decisions through the assessment of the relative importance of features concerning safety, ease of use and aspects of performance (Hilgerink et al. 2011; Hummel et al. 2012). Analytic hierarchy process (a form of MCDA) has also been used to predict the health economic performance of a new surgical technique and populate performance criteria for a diagnostic technology in development (Koning 2012). The MCDA process described by Koning established that the technology under development, an electronic nose for the diagnosis of infectious disease, was not sufficiently accurate or rapid to offer advantages in well-resourced settings but may be viable in lower-resource settings. Another use of MCDA has been to generate the criteria and key issues for two key stakeholder groups in the development of computer chips simulating organ functions for use in biotechnology development (Middelkamp et al. 2016).

1.7 What Are the Challenges in the Development and Assessment of Medical Devices?

1.7.1 What Are Medical Devices?

The term "medical device" includes a diverse range of medical technologies. The definition from the European Union device regulation (2017/745 2017) defines a medical device as:

> Any instrument, apparatus, appliance, software, implant, reagent, material or other article intended by the manufacturer to be used, alone or in combination, for human beings, for one or more of the following specific medical purposes: i) diagnosis, prevention, monitoring, prediction, prognosis, treatment or alleviation of disease; ii) diagnosis, monitoring, treatment, alleviation of or compensation for, an injury or disability, iii) investigation, replacement or modification of the anatomy or of a physiological or pathological process or state iv) providing information by means of in vitro examination of specimens derived from the human body, including organ, blood and tissue donations and which does not achieve its principal intended action by pharmacological, immunological or metabolic means, or in or on the human body, but which may be assisted in its function by such means. The following devices shall also be deemed to be medical devices: v) devices for the control or support of conception; vi) products specifically intended for the cleaning, disinfection or sterilisation of devices.

This definition is from EU regulations governing the licensing of medical devices (2017/745 2017). It is notable that the definition is quite wide but specifically excludes technologies that achieve their treatment effect by pharmacological, immunological, or metabolic means. This excludes drugs and vaccines as they have separate regulatory provisions. The U.S. Food and Drug Administration also regulates medical devices, and their definition is similar to the European definition.

Medical devices include several key categories such as large capital items, point of care devices, diagnostics, implantables, and wearable/digital devices. Examples of large capital items are a Da Vinci Robotic Surgery machine or a Magnetic Resonance Imaging (MRI) Scanner. These items have a large upfront capital cost and may be in use for a number of years. They may be used across different disease areas. Point of care (POC) devices take a service previously delivered in a laboratory setting, such as a blood test and deliver it at the point of care, which may be the bedside in a hospital ward or in a remote, rural healthcare center in an LMIC. Often POC tests are less accurate than laboratory tests but have the benefit of reaching more patients or being quicker with results. Diagnostics are tests that provide information and are broadly divided into in vivo diagnostics and in vitro diagnostics. In vivo diagnostics examine the body of the patient directly - for example, by monitoring blood pressure using a standard cuff. In vitro diagnostics take a sample of patient tissue or fluids and examine it in a laboratory environment - for example, the breast cancer prognostic test, Oncotype Dx, analyses a sample of tumor tissue in the laboratory and using an algorithm predicts whether or not a patient's breast cancer is likely to progress (ONCOTYPEIQ.COM). Implantables are devices that are directly implanted into the patient's body such as cardiac pacemakers. Wearable or digital devices are devices worn by, or implanted in, individual patients, which monitor symptoms or provide some therapeutic agent. An example of a wearable device is a subcutaneous insulin infusion or insulin pump that is implanted in a patient and delivers a constant level of insulin to a patient from an in-built reservoir.

1.7.2 Challenges Common to All medical Devices

1.7.2.1 Licensing and Regulation

Licensing and regulation of medical devices is highly complex and differs according to jurisdiction. The two main regulators of medical devices are the U.S. Food and Drug Administration (FDA) and the European Union (EU). Requirements for licensing in both Europe and the United States differ from those in place for pharmaceuticals. The level of evidence required for devices depends upon the classification of the device, hence the first hurdle for the developer is to understand which class their device is likely to fall into. In both the United States and the EU, devices other than in-vitro diagnostics are classified from Class I to Class III with Class III being the higher risk devices. For in-vitro diagnostics the categories are A-D.

In this paragraph, we use the EU to illustrate the complexity of licensing of medical devices. The United States has similarly complex requirements. In the EU, following classification of the medical device, it must undergo a conformity assessment, which involves the comparison of the device against essential requirements set out in the European Directives (EU 2017 745 and 746). The requirements cover issues such as whether the benefits outweigh the risks and whether the device achieves the claimed performance. Additional requirements cover chemical, physical, and biological properties, construction and environmental properties, potential contaminants, radiation protection, and the information to be supplied with the device such as instructions for use. The conformity assessment must include a clinical evaluation that demonstrates conformity with the relevant general safety and performance requirements, an evaluation of undesirable side-effects, acceptability of the benefit-risk ratio, and sufficient clinical evidence. The onus is on the manufacturer to decide and justify the level of clinical evidence necessary, which must be appropriate for the characteristics of the device and its intended use. At a minimum, a clinical evaluation must include a review of available clinical evidence in relation to the device in its intended use. Manufacturers must identify gaps in the evidence through a systematic literature review, which must be thorough and objective and include both favorable and unfavorable data. The depth and extent of the evaluation must be proportionate and appropriate to the nature, classification, intended purpose, and risks of the device. Novel and higher risk devices could be subject to additional clinical review by an expert panel. Clinical investigations must be performed for all implantable devices and Class III*.

Following successful conformity assessment, the device can receive a Conformité Européene (CE) Mark and be marketed anywhere in the EU. Manufacturers must undertake post-market surveillance and vigilance. Post-market surveillance is preventative and proactive, and charges the manufacturer with ensuring the ongoing safety of the device and that an appropriate risk-benefit balance is maintained. Vigilance is reactive and includes reporting of serious incidents with elements of voluntary and mandatory reporting.

Evidence requirements are lower, in both the EU and the United States, if it can be demonstrated that the new device is essentially the equivalent of an existing device. The Medical Device Regulation (EU 2017/745) has tightened up the definition of equivalence including three aspects, technical (design, specifications and physiochemical properties), biological (materials, duration of contact, release characteristics), and clinical (same stage or severity of disease, same site, same population, kind of user). It is also required to have contractual access to all technical documentation for the equivalent device on an ongoing basis for high risk devices and sufficient access for all other devices.

Class III devices require the clinical investigation to be a primary investigation if a claim of equivalence is not being made. In the United States, for non-equivalent devices falling into Class III and all implantables, primary clinical data is expected for the pre-market approval process. For equivalent devices, the lighter process is known as pre-market notification or the 510k process.

Evidence requirements for devices are not as well established as for pharmaceuticals, and it is advisable for developers to contact the regulators early in the development process for advice. In the UK, the Medicines and Healthcare Products Regulatory Agency, an executive agency of the Department of Health encourages manufacturers to contact the Innovation Office early in the development process to seek advice on regulatory requirements and manufacturing processes. It is also useful at this point to consider whether the evidence required by organizations who may adopt the device differs from that required for licensing in order that both sets of requirements are taken account of in designing the evidence generation plan.

1.7.2.2 Adoption

Once the device is licensed, the challenge for the developer is to secure sales in one or more jurisdictions. The relevant decision-maker will vary depending on the device and the context. For large capital equipment, the hospital will often be the decision-maker, whereas for a diagnostic test, the decision may be made at the regional or national level. For some devices there will be multiple layers of decision-makers. Some medical devices will be sold directly to patients or individual primary care practices. Each decision-maker may require a different level of evidence.

For example, a hypothetical diagnostic test is licensed according to EU regulations and can now be marketed in the UK. The developers can sell the diagnostic test directly to patients, but the market for this is likely to be small in a country with free, universal health care. The test could be sold to primary or secondary care providers directly, but they are unlikely to adopt it widely without the endorsement of the reimbursement agency, NICE. NICE has two separate assessment streams for diagnostics. The Diagnostics Assessment

Programme (DAP) assesses diagnostics that are likely to add cost and improve health. DAP uses methods similar to the mainstream Technology Assessment program for pharmaceuticals in that it compares the incremental benefit from using the test with the incremental cost and compares the consequent Incremental Cost-Effectiveness Ratio to a pre-established threshold. If the test offers value for money, it is likely to be recommended although not mandated (as a pharmaceutical would be after a successful TA process). The other stream of assessment is the Medical Technology Assessment Programme (MTAP), which assesses simpler technologies that offer cost savings compared to existing care. Again, if the diagnostic is recommended, it is not mandatory for the National Health Service (NHS) to adopt it although NICE do issue guidance to the NHS to try and encourage diffusion of innovation. However, due to contextual factors (such as what testing the local laboratory already undertakes, existing diagnostic and treatment pathways) each locality will need to assess whether the NICE conclusion applies to their context. Budgetary considerations are also likely to come into play where adoption of the device is not mandatory.

For some categories of device, additional guidance on the evidence required for recommendation may be available. This is the case for digital health technologies in the UK. In March 2019, NICE issued the Evidence Standards Framework for Digital Health Technologies (DHT) (NICE.ORG.UK) that sets standards of evidence for effectiveness and for economic evaluation. The Framework separates DHT into three tiers and sets the minimum and best practice evidence requirements for each tier. Tier 1 DHT is the lowest risk category and comprises devices that provide services to health and social care professionals and do not include any measurable patient outcomes. The evidence required to establish effectiveness involves demonstrating the credibility, reliability, and acceptability of the DHT with UK Health and Social Care professionals and evidence of accurate and reliable measurement and transmission of data, if applicable. The evidence requirements are cumulative in that higher tier devices need to fulfil all the requirements of lower tiers and their own requirements. For the highest-risk devices in Tier 3b (devices providing treatment or active monitoring of a patient's condition) the minimum required is a high-quality intervention study (either experimental or quasi-experimental) showing improvements in condition-specific outcomes or behaviors using a UK-relevant comparator. Again, recommendation by NICE on the provision of this evidence does not guarantee adoption, and further evidence may be required by local decision-makers.

1.7.2.3 Evidence

As we explained above, it is difficult to know what evidence is required for regulation and adoption of medical devices. Further difficulty arises in evidence generation due to the nature of some medical devices, which make randomized controlled trials (RCTs) difficult. For example, blinding is an issue in some trials of surgical devices. Although it is possible to set up trials using sham procedures, this adds complexity and expense to evidence generation. Recruitment in trials can also be difficult when invasive treatments are involved. Devices can be used as soon as they are licensed, and the evidence required to obtain a license sometimes falls short of the evidence some adopters require. As devices start to be used in some areas, developers may be disincentivised to generate evidence, or it may be more difficult to conduct controlled research if a device begins to replace its comparator technology. Policy-makers have a difficult balancing act as they do not want to delay the diffusion of a beneficial medical device by requiring onerous levels of evidence, but they need to ensure the safety of devices in the light of high-profile product recalls of breast implants and metal on metal hip replacements. Interestingly, the United States has higher evidence standards for licensing than Europe and also has less product recalls (Hwang et al. 2016). What is not clear is the opportunity cost of the delay in beneficial devices reaching the market.

1.7.3 Challenges Specific to Some Categories of Device

1.7.3.1 Learning Curve

Clinical outcomes often depend upon the training, competence, and experience of the clinician using the device. An RCT comparing a new device with a standard procedure may be demonstrating the difference between experience with the old procedure and inexperience with the new, rather than differences between the procedures themselves. This phenomenon is known as the learning curve. It needs to be taken account of in evidence generation to ensure that evidence from different centers and clinicians is comparable and reflects the potential of the device. Evidence generation strategies to avoid distortion by learning curve effects include setting trial criteria specifying a minimum level of previous operations, standardizing the level of experience of the centers, using statistical techniques to identify outcomes likely to be associated with learning curve or adjusting primary outcomes for a probable learning curve effect (Conroy et al. 2019; Papachristofi et al. 2016). A 2019 review of surgical trials in leading journals found that this is not routinely done and is poorly reported (Conroy et al. 2019).

1.7.3.2 Short Lifespan and Incremental Improvement

Medical devices are hard to evaluate as evidence is often generated on a range of models that are incrementally improved over a relatively short lifespan. Evidence may be generated on a range of devices that are being upgraded and changing capability. In addition to issues with clinical effectiveness evidence, prices may also change dramatically due to the market entry of new products, modifications, or volume discounts. Both factors mean that a technology appraisal could be outdated before it is completed.

1.7.3.3 Workflow

Many medical devices are dependent for their clinical efficacy and cost-effectiveness upon the context in which they are used and how they impact the current workflow. For example, an imaging device that was able to exclude malignant melanoma would only reduce the workload of specialist dermatologists in areas where those specialists currently saw all patients rather than in areas where some form of remote triage (for example, teledermatology) was already undertaken. Developers need to understand the workflow and user needs in all target settings in order to understand the clinical and economic value of the device. Devices that are disruptive, in the sense that they require re-engineering of workflows or investment in infrastructure, for example, will face greater barriers to diffusion.

1.7.3.4 Indirect Health Benefit

Diagnostic tests do not deliver health benefit by themselves. They improve health outcomes only if treatment changes because of the test result. As well as making evidence generation difficult, this results in cost savings potentially being realized in a different place in the clinical pathway. For example, part of the value proposition of Oncotype Dx, a commercially available prognostic test for patients with breast cancer is that chemotherapy costs will be avoided for a number of patients. Where the payer is an insurance company all incentives are aligned as the decision-maker that buys the test also pays for the chemotherapy. In other healthcare systems, the budget holder for diagnostic tests may be different from the drugs budget holder and incentives are not aligned. Awareness of the various incentives is essential for the developer in order to effectively market a diagnostic test.

1.7.3.5 Behavioral and Other Contextual Factors

The treatment effect of a medical device can be influenced by the preferences of both patients and clinicians. This makes generation of clinical evidence very challenging, as traditional randomized designs struggle to separate the impact of preferences and the treatment effect of the device. Potential solutions are innovative clinical study designs that take into account patients and/or clinicians' preferences (for example, the comprehensive cohort design (AbdElmagied et al. 2016) and qualitative studies. Developers of medical technologies must interact with potential users of their technology at an early stage of development so that they understand the impact of users' preferences.

1.7.3.6 Budgetary Challenges

As well as potential misalignment of incentives for diagnostic tests, other categories of device face budget challenges. Large capital items require upfront payment although developers may consider flexible business models that allow payment to be restructured. Implantable devices that offer longer lifespans also require investment upfront to save costs later, which may not be appealing to some payers.

1.8 The Contribution of DF-HTA in the Development and Translation of Medical Devices

It has been recognized internationally that the challenges faced by developers of medical devices are daunting. Delays in translation result in promising technologies not being available to improve the health of patients on a timely basis. In response to these challenges translational research bodies have been established with the aim of closing the translation gap. Two of the most active bodies in terms of published DF-HTA have been the Center for Translational Molecular Medicine (Steuten 2016) from the Netherlands and MATCH UK (MATCH.AC.UK 2018) a collaboration of UK universities. The following case studies are drawn from their portfolios with the exception of Kolominsky-Rabas et al. (Kolominsky-Rabas et al. 2015), which introduces the ProHTA project from Germany. The MaRS Excellence in Clinical Technology Evaluation (EXCITE) program in Canada (Levin 2015) is also worthy of note (MARSDD.COM 2019). MaRS EXCITE forms a model that European Healthcare Leaders are seeking to emulate where clinical trials are co-designed by developers and licensing and reimbursement authorities, and parallel submission to the authorities is encouraged (Ciani et al. 2018).

The case studies illustrate the role of DF-HTA in informing a broad set of decisions faced by developers of medical devices.

1.8.1 Case Study 1 - Identifying and Confirming Needs

This case study concerns an innovative laparoscopic instrument in development (Kluytmans et al. 2019). The developers initially thought that the instrument would improve various surgical procedures including meniscus surgery by replacing a number of different instruments in one, optimizing operating conditions for the surgeon and reducing the risk of infection thus improving patient experience and outcomes. The DF-HTA used a combination of qualitative stakeholder input to undertake a clinical value assessment followed by an economic value assessment using threshold and headroom analysis. The results of the qualitative exercise found that clinicians and other stakeholders, while enthusiastic about the new technology, felt that it did not offer significant incremental benefit over the current technology in the indication identified. A negative result is particularly valuable to developers as it allows them to refocus on a clinical area where there is more potential before significant expenditure has been incurred.

1.8.2 Case Study 2 - What Difference Could This Device Make?

ProHTA is an ambitious German project that uses a complex simulation model to assess technologies in development to see what clinical impact they may have and what characteristics the technology requires in order to deliver that performance. Kolominsky-Rabas et al. 2015 describe the application of the model in assessing Mobile Stroke Units. The units offer clinical value as they potentially reduce time to thrombolysis, thus improving patient outcomes. The simulation model demonstrated that the mobile stroke units save up to 49 minutes between the ambulance call and the therapy decision, and that this is enough to move 16.4 percent of patients into the most favorable time interval for thrombolysis.

1.8.3 Case Study 3 - Which Research Project Has the Most Potential?

This case study reports a multi-criteria decision analysis (MCDA) approach to clinical and economic value assessment in order to prioritize potential research projects in CTMM (de Graaf et al. 2015). The goal of the MCDA was to identify the biomarker project with the most potential to reduce the burden of Type 2 diabetes. The first step involved the identification of six criteria for the assessment from discussion with decision-makers. Criteria included three relating to barriers to realize potential and three concerning cost and impact on quality of life. The second step involves scoring each alternative project on the six criteria (either numerically, where this can be estimated or ordinally). The third step calculates the preferred option, making different assumptions about the weighting of the different criteria. The process allowed the developers to exclude one of the four research options regardless of decision-maker preferences and to rank the other three dependent upon whether the decision-maker prioritizes financial criteria or ease of implementation.

1.8.4 Case Study 4 - What Is the Required Performance to Deliver Clinical Utility?

In another case study from CTMM, Buisman et al. (2016) undertook a clinical and economic value assessment of four diagnostic tests for rheumatoid arthritis (B-cell test, IL6 test, imaging of hands and feet, and genetic assay). The tests were assessed using a five-year health economic model in the form of a decision tree followed by an individual-level state-transition model. The study illustrates the complexity of diagnostic assessment as it models each test in two different places in the current diagnostic pathway and in different sub-populations based on the risk of RA. There was much uncertainty around costs and estimates of test performances, as is usually the case in DF-HTA, but the study was able to identify a likely dominant strategy (the B-cell test) and determine the cost and test performance required for the preferred option to be cost-effective (EUR 200-300. High specificity).

1.8.5 Case Study 5 - What Are the Key Parameters for Evidence Generation?

Vallejo-Torres et al. (2011) carried out a three-stage economic evaluation of absorbable pins for Hallux Valgus compared to sutures and metallic pins. The early stage used headroom analysis to prioritize competing products, the mid-stage began to synthesize available data into a decision model, and the final stage developed a full decision model. The mid-stage model allowed the exploration of uncertainty and the identification (through Value of Information analysis) of those parameters for which further information would be most valuable. The study found that quality of life data was the most valuable data to collect in the next stage of evidence generation.

1.9 Conclusion

Although engineering and technology have the potential to revolutionize patient care and health outcomes, there is evidence of much waste in healthcare research (Greenhalgh et al. 2018). There is a tendency for research to be technology led rather than needs led (Kluytmans et al. 2019) and evidence that funding for enterprising firms in health technology is not always aligned with health research funding streams (Lehoux et al. 2017). Evidence suggests that many developers of medical devices do not use formal methods of decision-support and do not have in-house capacity and knowledge to perform HTA (Craven et al. 2012; Markiewicz et al. 2017). This, despite the immense challenges faced by developers of medical technologies, particularly in evidence development, licensing, regulation, and adoption.

The case studies described above suggest that there are benefits for developers in consulting an HTA practitioner at an early stage in the development process. The most immediate benefit is that the HTA is able to inform a go/no-go decision at any point in development. The first case study demonstrated that there was little potential value in pursuing the project if the target was meniscus surgery. This negative finding is extremely useful to developers, who can then focus their efforts in other clinical areas. Where the indication was that the project had potential clinical utility, as in the other four case studies, DF-HTA offers benefits. In the second case study, the simulation model demonstrated to external stakeholders the potential utility of the Mobile Stroke Units in the specific context of the German healthcare system. The third case study allowed the exploration of decision-makers' criteria and understanding of the relative priority of the projects dependent upon the weighting between them. It also allowed the rejection of one of the alternatives regardless of the weighting between criteria. This prioritization took place before significant expenditure on any of the biomarker projects illustrating the usefulness of DF-HTA in avoiding research waste. The fourth case study illustrates the complexity of a contextual analysis of diagnostic tests that would be difficult for many developers to undertake without the support of a translational research body like CTMM. The useful information generated from this exercise allowed the prioritization of the B-cell test, which dominated in most modelled scenarios and provided the developers with guidelines as to the performance and test costs required in order to make the test cost-effective in the Netherlands. The final case study demonstrated the iterative nature of DF-HTA and showed how earlier stages of analysis can guide evidence development in later stages, indicating which are the most important parameters for cost-effectiveness.

DF-HTA has a role to play in maximizing the potential of development in science and technology. DF-HTA can help ensure that medical devices in development meet a clinical need and deliver clinical utility in the context in which it is hoped they will be used. They can alert developers at an early stage where a device is unlikely to make a difference and provide information about how good the device has to be to make a difference. Undertaken early in the device development process, evidence generation can be tailored to ensure that it meets the needs of licensing and reimbursement agencies. As is clear from the case studies, translational research bodies that facilitate links between commercial entities, academic and clinical researchers, clinicians, regulators, and reimbursement agencies, have a role to play in the provision of DF-HTA. This, in turn, may improve the translation rate, ensuring that patients receive timely access to innovative engineering and technology solutions.

References

EU Medical Devices Regulation 2017/745, 2017. URL https://eur-lex.europa.eu/legal-content/EN/TXT/?uri=CELEX\LY1\textbackslash%3A32017R0745.

AbdElmagied, A,M., L.E. Vaughan, A.L. Weaver, S.K. Laughlin-Tommaso, G.K. Hesley, D.A. Woodrum, V.L. Jacoby, M.P. Kohi, T.M. Price, and A. Nieves, "Fibroid interventions: reducing symptoms today and tomorrow: extending generalizability by using a comprehensive cohort design with a randomized controlled trial," *American Journal of Obstetrics and Gynecology*, 215(3), p. 338. e1–338. e18, 2016. ISSN 0002–9378.

Bennette, C.S., D.L. Veenstra, A. Basu, L.H. Baker, S.D. Ramsey, and J.J. Carlson, "Development and evaluation of an approach to using value of information analyses for real-time prioritization decisions within SWOG, a large cancer clinical trials cooperative group," *Medical Decision Making*, 36 (5), p 641–651, 2016. ISSN 0272-989X.

Breteler, M. (2012). "Scenario analysis and real options modeling of home brain monitoring in epilepsy patients," doctoral dissertation, University of Twente, Enschede, the Netherlands.

Buisman, L.R., J.J. Luime, M. Oppe, J.M.W. Hazes, and M.P.M.H. Rutten-van Mlken. (2016). "A five-year model to assess the early cost-effectiveness of new diagnostic tests in the early diagnosis of rheumatoid arthritis," *Arthritis Research & Therapy*, 18(1), p135, ISSN 1478-6362.

Carlson, J.J., R. Thariani, J. Roth, J. Gralow, N.L. Henry, L. Esmail, P. Deverka, S.D. Ramsey, L. Baker, and D.L. Veenstra. (2013) "Value-of-information analysis within a stakeholder-driven research prioritization process in a US setting: an application in cancer genomics. *Medical Decision Making*, 33(4), p463–471, ISSN 0272-989X.

Amanda Megan Chapman. (2013) "The use of early economic evaluation to inform medical device decisions: an evaluation of the headroom method," doctoral dissertation, University of Birmingham, Birmingham, UK.

Ciani O., Federici C., Pecchia L. (2018) *The evaluation of medical devices: are we getting closer to solve the puzzle? A review of recent trends.* In: Eskola H., Väisänen O., Viik J., Hyttinen J. (eds) EMBEC & NBC 2017. EMBEC 2017, NBC 2017. IFMBE Proceedings, vol 65. Springer, Singapore. https://doi.org/10.1007/978-981-10-5122-7_229.

Claxton, K., P. Revill, M. Sculpher, T. Wilkinson, J. Cairns, and A. Briggs. (2014) The Gates reference case for economic evaluation. Seattle, WA: The Bill and Melinda Gates Foundation.

Cochrane, A.L. (1972) Effectiveness and efficiency: random reflections on health services. London: Nuffield Provincial Hospitals Trust. London: Nuffield Provincial Hospitals Trust, London.

Conroy, E.J., A. Rosala-Hallas, J.M. Blazeby, G. Burnside, J.A. Cook, and C. Gamble. (2019) "Randomized trials involving surgery did not routinely report considerations of learning and clustering effects," *Journal of Clinical Epidemiology*, p107, P27–35, ISSN 0895-4356.

Craven, M.P., M.J. Allsop, S.P. Morgan, and J.L. Martin, (2012) "Engaging with economic evaluation methods: insights from small and medium enterprises in the UK medical devices industry after training workshops," *Health Research Policy and Systems*, 10(1):29, ISSN 1478-4505.

Davey, S.M. M. Brennan, B.J. Meenan, R. McAdam, A. Girling, A. Chapman, and R.L. Lilford. (2011) "A framework to manage the early value proposition of emerging healthcare technologies," *Irish Journal of Management*, 31(1), ISSN 1649-248X.

de Graaf, G., D. Postmus, and E. Buskens. (2015) "Using multicriteria decision analysis to support research priority setting in biomedical translational research projects," *BioMed Research International*, ISSN 2314-6133.

Drummond, M.F., M.J. Sculpher, K. Claxton, G.L. Stoddart, and G.W. Torrance. (2015) *Methods for the economic evaluation of health care programmes*. Oxford: Oxford University Press, ISBN 0191643580.

Eddy, D. (2009) "Health technology assessment and evidence-based medicine: what are we talking about?" *Value in Health*, 12: pS6–S7, ISSN 1098-3015.

Girling, A.J. G. Freeman, J.P. Gordon, P. Poole-Wilson, D.A. Scott, and R.J. Lilford. (2007) "Modeling payback from research into the efficacy of left-ventricular assist devices as destination therapy," *International Journal of Technology Assessment in Health Care*, 23(2), p 269–277, ISSN 1471-6348.

Gosling, J.P. (2014) "Methods for eliciting expert opinion to inform health technology assessment," Vignette Commissioned by the MRC Methodology Advisory Group. Medical Research Council (MRC) and National Institute for Health Research (NIHR).

Gosling, J.P. (2018) *SHELF: the Sheffield elicitation framework*, Springer, p61–93.

Greenhalgh, T. N. Fahy, and S. Shaw. (2018) "The bright elusive butterfly of value in health technology development; Comment on 'Providing value to new health technology: The early contribution of entrepreneurs, investors, and regulatory agencies,' " *International Journal of Health Policy and Management*, 7(1), p81.

Haakma, W., L. Bojke, L. Steuten, and M. IJzerman. (2011) "Pmd4 expert elicitation to populate early health economic models of medical diagnostic devices in development," *Value in Health*, 14(7), pA244–A245, ISSN 1098-3015.

Hilgerink, M.P., M.J.M. Hummel, S. Manohar, S.R. Vaartjes, and M.J. IJzerman. (2011) "Assessment of the added value of the Twente photoacoustic mammoscope in breast cancer diagnosis," *Medical Devices (Auckland, NZ)*, 4, p107–15.

Hummel, J.M., I.S.M. Boomkamp, L.M.G. Steuten, B.G.J. Verkerke, and M.J. Ijzerman. (2012) "Predicting the health economic performance of new non-fusion surgery in adolescent idiopathic scoliosis," *Journal of Orthopaedic Research*, 30(9), p1453–1458, ISSN 0736-0266.

Hwang, T.J., E. Sokolov, J.M. Franklin, and A.S. Kesselheim. (2016) "Comparison of rates of safety issues and reporting of trial outcomes for medical devices approved in the European Union and United States: cohort study," *BMJ*, 353:i3323, ISSN 1756-1833.

ICER.ORG (2019) URL HTTPS://ICER-REVIEW.ORG/ABOUT/SUPPORT/.

Iglesias, C.P., A. Thompson, W.H. Rogowski, and K. Payne (2016) "Reporting guidelines for the use of expert judgement in model-based economic evaluations," *Pharmacoeconomics*, 34(11), p1161–1172, ISSN 1170-7690.

Ijzerman M.J., and L.M.G. Steuten. (2011) "Early assessment of medical technologies to inform product development and market access," *Applied Health Economics and Health Policy*, 9(5), p331–347, ISSN 1175-5652.

Ijzerman, M.J., H. Koffijberg, E. Fenwick, and M. Krahn. (2017) "Emerging use of early health technology assessment in medical product development: a scoping review of the literature," *Pharmacoeconomics*, 35(7), p727–740, 2017. ISSN 1170-7690.

INAHTA.ORG, 2019. URL HTTP://WWW.INAHTA.ORG/.

INTUITIVE.COM, 2019. URL https://www.intuitive.com/en-us/products-and-services/da-vinci.

Joosten, S.EP., V.P. Retel, V.M.H. Coup, M.M. van den Heuvel, and W.H. Van Harten. (2016) "Scenario drafting for early technology assessment of next generation sequencing in clinical oncology," *BMC Cancer*, 16(1), p66, ISSN 1471-2407.

Kip, M.M.A., L.M.G. Steuten, H. Koffijberg, M.J. Ijzerman, and R. Kusters. (2018) "Using expert elicitation to estimate the potential impact of improved diagnostic performance of laboratory tests: a case study on rapid discharge of suspected non–ST elevation myocardial infarction patients," *Journal of Evaluation in Clinical Practice*, 24(1), p31–41, ISSN 1356-1294.

Kluytmans, A., M. Tummers, G.J. van der Wilt, and J. Grutters. (2019) "Early assessment of proof-of-problem to guide health innovation," *Value in Health*, 22(5), p601–606, ISSN 1098-3015.

Kolominsky-Rabas, P.L., A. Djanatliev, P. Wahlster, P. Gantner-Br, B. Hofmann, R. German, M. Sedlmayr, E. Reinhardt, J. Schttler, and C. Kriza. (2015) "Technology foresight for medical device development through hybrid simulation: The ProHTA project," *Technological Forecasting and Social Change*, 97, p105–114, ISSN 0040-1625.

Koning, R. (2012), Bacterial pathogen identification in blood samples with electronic nose technology in a clinical setting: an early medical technology assessment of the µDtect," Masters thesis, University of Twente, Enschede, the Netherlands.

Lehoux, P., F.A. Miller, G. Daudelin, and J.-L. Denis. (2017) "Providing value to new health technology: the early contribution of entrepreneurs, investors, and regulatory agencies," *International Journal of Health Policy and Management*, 6(9), p509.

Levin, L. (2015) Early evaluation of new health technologies: the case for premarket studies that harmonize regulatory and coverage perspectives," *International Journal of Technology Assessment in Health Care*, 31(4), p.207–209, ISSN 0266-4623.

LYGATURE.ORG, 2019. URL HTTPS://WWW.LYGATURE.ORG/CTMM-PORTFOLIO.

Markiewicz, K., J. Van Til, and M. IJzerman. (2017), "Early assessment of medical devices in development for company decision making: an exploration of best practices," *Journal of Commercial Biotechnology*, 23(2), ISSN 1462-8732.

MARSDD.COM. (2019) URL HTTPS://WWW.MARSDD.COM/SYSTEMS-CHANGE/MARS-EXCITE/MARS-EXCITE/.

MATCH.AC.UK. (2018). URL HTTP://WWW.MATCH.AC.UK.

Meltzer, D.O., T. Hoomans, J.W. Chung, and A. Basu. (2011) "Minimal modeling approaches to value of information analysis for health research," *Medical Decision Making*, 31(6), pE1–E22, ISSN 0272-989X.

Middelkamp, H.H.T., A.D. van der Meer, J.M. Hummel, D.F. Stamatialis, C.L. Mummery, R. Passier, and M.J. Ijzerman. (2016) "Organs-on-chips in drug development: the importance of involving stakeholders in early health technology assessment," *Applied in Vitro Toxicology*, 2(2), p.74–81, ISSN 2332-1512.

Morris, D.E., J.E. Oakley, and J.A. Crowe. (2014) "A web-based tool for eliciting probability distributions from experts," *Environmental Modelling & Software*, 52, p. 1–4, 2014. ISSN 1364-8152.

NICE.ORG.UK. URL HTTPS://WWW.NICE.ORG.UK/ABOUT/WHAT-WE-DO/OUR-PROGRAMMES/EVIDENCE-STANDARDS-FRAMEWORK-FOR-DIGITAL-HEALTH-TECHNOLOGIES.

ONCOTYPEIQ.COM. URL https://www.oncotypeiq.com/en-gb/breast-cancer/healthcare-professionals/oncotype-dx-breast-recurrence-score/about-the-test.

Ostergaard, S., and C. Mldrup. (2010) "Anticipated outcomes from introduction of 5-httlpr genotyping for depressed patients: an expert Delphi analysis," *Public Health Genomics*, 13(7-8), p. 406–414, ISSN 1662-4246.

Papachristofi, O., D. Jenkins, and L.D. Sharples. (2016) "Assessment of learning curves in complex surgical interventions: a consecutive case-series study," *Trials*, 17(1), p. 266, ISSN 1745-6215.

SOCIALVALUEUK.ORG. SROI Guide. 2019. URL http://www.socialvalueuk.org/resources/sroi-guide/.

Steuten, L.M., (2016) "Multi-dimensional impact of the public–private center for translational molecular medicine (ctmm) in the Netherlands: understanding new 21st century institutional designs to support innovation-in-society," *Omics: a Journal of Integrative Biology*, 20 (5), p. 265–273, ISSN 1536-2310.

Stevens, A., R. Milne, and A. Burls. (2003) "Health technology assessment: history and demand," *Journal of Public Health*, 25(2), p. 98–101, ISSN 1741-3850.

Terjesen, C.L., J. Kovaleva, and L. Ehlers. (2017) "Early assessment of the likely cost effectiveness of single-use flexible video bronchoscopes," *PharmacoEconomics-Open*, 1(2), p. 133–141, ISSN 2509-4262.

Vallejo-Torres, L., L. Steuten, B. Parkinson, A.J. Girling, and M.J. Buxton. (2011) "Integrating health economics into the product development cycle: a case study of absorbable pins for treating hallux valgus," *Medical Decision Making*, 31(4): 596–610, ISSN 0272-989X.

Wahlster, P., M. Goetghebeur, C. Kriza, C. Niederlnder, and P. Kolominsky-Rabas. (2015) "Balancing costs and benefits at different stages of medical innovation: a systematic review of multi-criteria decision analysis (mcda)," *BMC Health Services Research*, 15(1), p. 262.

WHO.INT, 2019. URL HTTPS://WWW.WHO.INT/HEALTH-TECHNOLOGY-ASSESSMENT/ABOUT/DEFINING/EN/.

Wong, W.B., S.D. Ramsey, W.E. Barlow, L.P. Garrison Jr, and D.L. Veenstra. (2012) "The value of comparative effectiveness research: projected return on investment of the rxponder trial (swog s1007),"*Contemporary clinical trials*, 33(6), p. 1117–1123.

2

Contactless Radar Sensing for Health Monitoring

Francesco Fioranelli[1] *and Julien Le Kernec*[2]

[1] *Microwave Sensing Signals and Systems (MS3) section, Department of Microelectronics, at the Delft University of Technology*
[2] *James Watt School of Engineering, University of Glasgow*

Contactless radar and radio frequency (RF) sensing has recently gained much interest in the domain of health care and assisted living, due to its capability to monitor relevant parameters for the health and well-being of people. Applications range from the monitoring of respiration and heartbeat to mobility levels, gait and locomotion parameters, and behavioral patterns. What makes radar sensing attractive compared to alternative and complementary technologies such as video-cameras, wearables, or ambient sensors, is their contactless capabilities, whereby no sensors need to be worn by the people monitored, and no plain images or videos need to be collected in private spaces and homes.

The aim of this chapter is to provide an overview of recent radar and RF technologies applied to the domain of health care, demonstrating the relevant information obtainable from the transmission, propagation, and reception of RF signals and their interaction with people in their daily environment.

2.1 Introduction: Healthcare Provision and Radar Technology

In recent years, new needs in welfare and healthcare provision have emerged from the rapidly aging population worldwide, specifically in all Western countries but also in China. Statistics from the World Health Organisation and the United Nations estimate that about 30% of the world population will be older than 65 years by 2050; in the United Kingdom, the Office for National Statistics expects that the proportion of people older than 85 years will double in the next 20 years (Storey 2018), and very similar estimates exist across a large number of countries.

Non-communicable diseases (NCDs) such as coronary heart disease, diabetes, cancer, and chronic respiratory disease significantly increase their incidence with aging, and individuals can be affected by multiple NCDs (multimorbidity); furthermore, aging also increases the likelihood of critical events such as stroke and dramatic falls (Barnett et al. 2012; Rijken et al. 2018). Effectively managing these conditions is challenging and requires

Engineering and Technology for Healthcare, First Edition.
Edited by Muhammad Ali Imran, Rami Ghannam and Qammer H. Abbasi.

monitoring over long periods of time a large number of vital signs and biomarkers to create a comprehensive picture of the health condition of the people affected. Current health systems are typically organized to respond to a single condition per individual, with intensive care provided in a specialized hospital structure. However, increasing numbers of potential patients combined with budget pressure on public funding risk making this model unsustainable. Beyond economic sustainability, the quality of life of patients and families are affected by prolonged periods spent in hospitals, with the additional risk of exposure to antibiotic-resistant bacteria.

In this context, the use of the most advanced technologies to provide integrated care in private home environments has attracted significant interest (Aykan 2012). The objectives are twofold. First, hospitalization should be avoided and resorted to only for the most critical situations, leveraging the new technologies' capabilities to monitor the health care of vulnerable people in their homes. This would enable the preservation of their autonomy and independence, preventing disruptions to daily routine and separation from familiar environments that can have a significant impact on people's well-being. This aspect applies not only to reducing admissions to hospitals, but also to expediting discharges, as in-home technological solutions can support required monitoring and rehabilitation in the patients' own homes. The second objective is to use technology to shift the paradigm from "reactive health care" intervening after a problem has happened, to "proactive health care" that exploits the continuous monitoring capabilities enabled by new technologies for timely identification of any sign related to worsening health conditions. This would enable timely and personalized treatment for health conditions before their effects become too significant and disruptive for the people affected.

Why should radar/RF sensing be considered as suitable technologies in ambient assisted living? Many technologies have been proposed (Debes et al. 2016; Chacour et al. 2017; Cippitelli et al. 2017) and all can be evaluated based on different metrics, including accuracy and quality of the information provided, reliability, cost, and perception/acceptance from the end-users. Technology acceptance is particularly significant as the technology needs to be deployed in private home environments (e.g., bedrooms). Compared with alternative technologies, radar sensing can present interesting advantages:

- Unlike the situation with wearable sensors, the users are not required to change their daily routine and wear, carry, or interact with additional electronic devices. This freedom can support the acceptance of the technology and is more suited for cognitively impaired elderly/patients.
- Unlike cameras, no plain videos or images of the users and their homes are collected when using radar or RF systems, possibly reducing the perceived impact in terms of privacy. Furthermore, from the perspective of cyber-security and potential hacking of the sensors with uncontrolled and unauthorized disclosure of their data, radar data or images are more difficult to interpret than videos or images, making their disclosure potentially less damaging;
- Unlike ambient sensors (such as pressure, presence, on-off switches), there is no need to retrofit existing homes with many additional new devices, but the radar/RF system can be a simple and compact box similar to a Wi-Fi router, or the Wi-Fi router itself in the case of sensing based on passive radar framework or Wi-Fi channel state information.

Figure 2.1 shows the concept of the emerging radar-sensing technologies in a home for daily activity recognition, vital signs checking, and long-term health monitoring

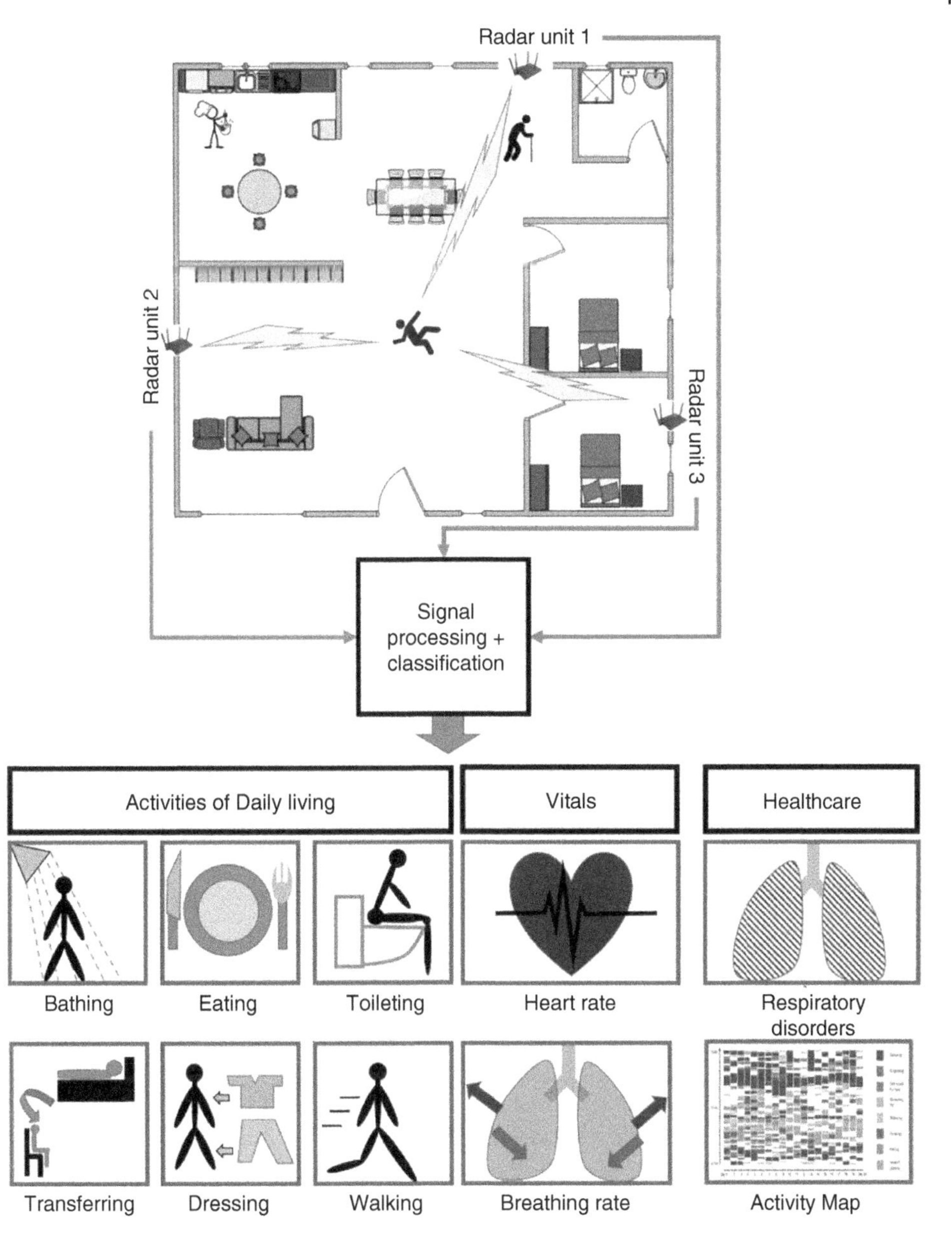

Figure 2.1 In-home sketch scenario of radar-based sensing technologies for health care. Source: J. Le Kernec et al., © 2019, IEEE.

applications. Multiple radar sensors, as shown in the top part of Figure 2.1, can provide additional, spatially diverse information for better monitoring, at the expense of increased cost and complexity of the system. Through-wall penetration and imaging capabilities of radar sensing could also be exploited for enhanced coverage. The number and deployment locations of radar-sensing nodes to optimally monitor a given environment is still an outstanding research question. A growing body of literature shows the usage of radar sensing to monitor a number of vital signs and biomarkers (Boris-Lubecke et al. 2016), for example, respiration (Adib et al. 2015; Chen et al. 2018), heart rate and related parameters of cardiac functionalities (Sakamoto and Yamashita 2019; Dong et al. 2019; Wu et al. 2019) blood pressure (Kuwahara, Yavari, and Boric-Lubecke, 2019; Ebrahim et al., 2019; Ohata, Ishibashi, and Sun. 2019; Buxi, Redouté, and Yuce, 2017) glucose concentration relevant for diabetes (Omer et al. 2018; Lee et al. 2018). This volume of research on usage of radar sensing shows the potential to use this tool to monitor NCDs in home environments, that is, to monitor over time the variety of vital signs and biomarkers necessary to understand the complex health situation of vulnerable subjects and its evolution. The remainder of this chapter will provide some basics on radar principles relevant for these applications a non-exhaustive list of examples of results. The chapter is organized as follows. Section 2.2 describes some fundamentals about radar sensing, with signal processing and machine learning typically applied in the context of health care. Section 2.3 presents a comprehensive set of results from recent research works on various applications of radar and RF sensing, from the recognition of activities and falls, to the monitoring of gait and vital signs. Section 2.4 concludes with a discussion of outstanding research challenges and trends on possible future developments.

2.2 Radar and Radar Data Fundamentals

This section summarizes the fundamental principles of radar systems and sensing, aiming to provide readers with a synthetic background on this technology. For additional details and comprehensive explanations, readers are referred to the referenced books (Skolnik 2008; Chen, Tahmoush, and Miceli 2018; Amin 2018).

2.2.1 Principles of Radar Systems

The fundamental principle of any radar system consists of transmitting and receiving sequences of electromagnetic waves, whose amplitude, frequency, and/or phase are modulated in suitable waveforms. By collecting and analyzing the received radar waveforms, it is possible to extract information on targets of interest in the area under test from the transmitted EM waves that are reflected back to the receiver. Hence, in its most basic model, a radar system will have a transmitter and a receiver block to generate, condition, and capture electromagnetic waves; antennas for transmission and reception of the radar waveforms; and a digital unit to process, store, and visualize the radar data.

Two main families of radar waveforms are used for sensing in the healthcare context, pulsed waveforms and continuous wave (CW) waveforms. Pulsed radar systems transmit pulses with a given periodicity set by the Pulse Repetition Frequency (PRF) parameter and receive replicas of the transmitted pulses reflected back by objects in the environment. The range (physical distance) R to a target can, therefore, be calculated as in equation (2.1), where c is the speed of light and τ is the time delay for the radar pulse to travel from the transmitter to the object and back to the receiver.

$$R = \frac{c\tau}{2} \tag{2.1}$$

However, for the presence of an object to be detected by a radar system, the reflected pulses need to contain enough electromagnetic energy with respect to the internal noise of the radar system. The power budget of any radar system can be represented by the so-called radar equation, which is reported in equation (2.2) in its basic form. P_r and P_t are the received power and transmitted power of the radar, respectively, G_t and G_r are the gain of the transmitter and receiver antennas respectively, σ is the Radar Cross Section (RCS, measured in square meters) parameter that models how well a certain target reflects radar waveforms depending on its shape and materials, and λ is the wavelength of the radar waveform (Skolnik 2008; Amin 2018). The radar equation can be reformulated to make explicit reference to the Signal to Noise Ratio (SNR) necessary at the radar receiver for proper detection and processing of the objects' echoes, essentially for the radar to work with acceptable performances. Equation (2.3) shows the maximum detection range R_{max} for a given target (represented by its RCS) and a given radar system (represented by its required SNR), where K_b is Boltzmann's constant, T_0 is the temperature of the radar in Kelvin, B is the noise bandwidth of the radar receiver, and N_F is the noise figure of the radar receiver.

$$P_r = \frac{P_t G_t G_r \sigma \lambda^2}{R^4 (4\pi)^3} \tag{2.2}$$

$$R_{max} = \sqrt[4]{\frac{P_t G_t G_r \sigma \lambda^2}{(4\pi)^3 K_b T_0 B N_F \,(SNR)}} \tag{2.3}$$

Equations (2–3) have an important practical implication for the choice of frequency for the radar waveforms. The received power and therefore the maximum range for detection increase with longer wavelengths and lower operating frequencies; hence, from the perspective of the power budget, lower frequency appears to be favorable.

Pulsed radar systems typically use very narrow pulses with a low duty cycle, that is, short pulses in terms of their temporal duration with respect to the pulse repetition period; this generates wide bandwidths in terms of the frequency domain representation of the radar waveforms, often Ultra Wide Bands (UWB systems). Such wide bandwidth is an advantage for providing adequate range resolution, that is, the minimum distance between two closely located objects to be perceived as two separated targets. The range resolution ΔR can be calculated as in equation (2.4), where T_p is the temporal duration of the radar pulse and B its bandwidth.

$$\Delta R = \frac{cT_p}{2} = \frac{c}{2B} \tag{2.4}$$

However, a disadvantage of using short pulses is the requirement of high transmit peak power to achieve the necessary average transmit power for the detection of objects of interest. The challenge in achieving this requirement with acceptable complexity and cost of the relevant electronics components led to the development of CW radar systems and their modulated version, FMCW (frequency modulated continuous wave). FMCW radar systems transmit and receive continuous waveforms, where the instantaneous transmitted frequency is accordingly modulated over time, for example, with triangular, saw-tooth, quadratic, hyperbolic waveforms. The FMCW signal is often called the "chirp" signal. The range (physical distance) to an object is proportional to the difference in frequency between the transmitted and the received FMCW signals, which is typically referred to as the "beat frequency" and can be obtained at the radar receiver by mixing the received FMCW signal with a replica of the transmitted signal (Stove 1992). Equation (2.5) shows how the range to a given object is proportional to the time delay τ (the time for the radar waveform to propagate from the transmitter to the object and be reflected back to the receiver) as for a pulsed radar, but then the time delay is proportional to the beat frequency f_b through the period of the FMCW waveform T (chirp duration) and the maximum frequency deviation B (the bandwidth of the chirp between its minimum and maximum instantaneous frequency).

$$R = \frac{c\tau}{2} = \frac{cf_b T}{2B} \tag{2.5}$$

Note that for indoor environments the maximum range to be covered is limited to typically 10–20m maximum; therefore we can use higher carrier frequencies to miniaturize the radar frontend and access wide bandwidth for increased range resolution.

Besides measuring the range to an object, radar systems can also use the Doppler effect to measure the velocity at which such an object may be moving. It is a change in the frequency (Doppler frequency shift) of the received radar waveforms due to movement of the object. The Doppler shift f_d can be calculated as in equation (2.6), where v is the velocity of the object, f_c is the operating frequency of the radar waveform, and θ is the aspect angle, that is, the angle between the line of sight of the radar and the direction of the object's velocity vector.

$$f_d = -\frac{2vf_c}{c}\cos(\theta) \tag{2.6}$$

Equation 6 shows that the Doppler shift is minimal (ideally zero) when the aspect angle is close to 90° or 270°, that is, when the object is moving along a tangential trajectory with respect to the radar line of sight, and vice versa maximal when the target trajectory is radial, that is, the object is heading straight towards the radar or away from it. The presence of the cosine can change the sign of the Doppler shift, to make it positive when the object is heading towards the radar, and negative when moving away. Furthermore, the perceived Doppler shift is larger, hence easier to measure, if the operating frequency of the radar is higher. This can be a conflicting requirement with the comment for equation (2.3) as to the preference of lower operational frequency for better power budget and longer detection range. A trade-off between maximum range and the amplitude of Doppler frequency shifts to more effectively discriminate with larger shifts against the echoes of static objects (in technical terms referred to as "clutter") whose Doppler frequency shift is close to 0 Hz.

The measurement of the Doppler shift and therefore of the velocity of an object of interest is typically performed looking at the changes in frequencies across a set of received radar waveforms over time. This approach could be interpreted as a "sampling problem," where the unknown quantity (the Doppler shift) needs to be sampled by at least twice its maximum expected value to comply with Nyquist sampling theorem, where taking a sample in this context means sending and receiving a radar waveform. In other words, the faster the target is expected to be, the higher its induced Doppler frequency shift, hence the faster the radar will be required to transmit and receive waveforms to sample this signal without ambiguity. This leads to equation (2.7), which links the maximum unambiguous Doppler shift measurable by a radar system $f_{d,max}$ with its Pulse Repetition Frequency *PRF* (for pulsed radar systems) or the chirp duration T (for FMCW radar systems). Note that the unambiguous range is symmetrical across positive and negative values, depending on movements toward the radar (positive) or away (negative).

$$f_{d,max} = \pm\frac{PRF}{2} = \pm\frac{1}{2T} \tag{2.7}$$

2.2.2 Principles of Radar Signal Processing for Health Applications

Broadly speaking, the two main applications of radar sensing in the healthcare context are the classification of "macro human activities," such as walking, sitting, standing, bending, moving objects, and identification and monitoring of vital signs such as respiration, heart rate, blood pressure, and cardiac activity, as mentioned in the Introduction. Both are based on the detection and characterization of movements of body parts of the subject, either larger movements of the whole body and limbs (for the macro activities) or very small movements of the chest and abdomen (for respiration monitoring), limbs (for tremors monitoring), and internal organs (for heartbeat monitoring or blood pressure estimation).

Once the presence of a subject is noticed, the typical radar signal processing on the data aims to characterize their movements in three domains: range (as the distance at which the subject and his or her body parts are located with respect to the radar), time (as the evolution over time of the position of the subject and any change to his or her movements), and velocity (as the speed at which these changes happen, whether controlled and regular or with sudden acceleration and deceleration). The previous section has discussed how measuring velocity in radar terms means calculating the frequency components of the received waveforms, specifically any changes induced by the Doppler effect. In terms of radar signal processing, this frequency characterization can be done using Fourier analysis implemented by the fast Fourier transform (FFT) algorithm.

Figure 2.2 shows examples of radar signatures in pairs of the aforementioned three domains range-time-Doppler for the simple case of a person walking away from the radar, in this case, an ultra-wideband (UWB) pulsed radar system operating in the X-band. The first format for the signatures is the Range-Time matrix, where the received radar pulses are stacked together one below the other, so that the vertical dimension represents the sequence of pulses associated with time, and the horizontal dimension the digitized samples related to the physical distance of possible objects of interest. In this case, as the person was walking away from the radar, the Range-Time (RT) image shows a diagonal pattern with the signature located at increasing range over time. A single FFT operation can be applied across the time dimension of the RT image (by column in Figure 2.2) to obtain a Range-Doppler matrix that characterizes the overall movement of the object. In this case, the Doppler contribution is negative, as the person was walking away from the radar; one can note that the Doppler shift (velocity) is low just below 1m range (very

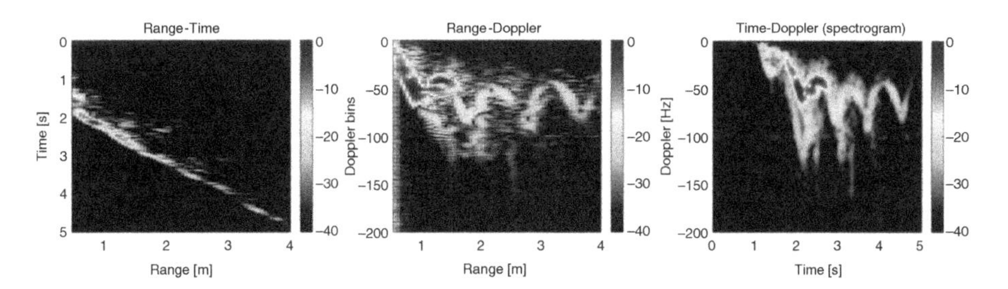

Figure 2.2 Example of radar data format in the three main domains of range (distance), time, and Doppler (velocity) for a simple example of a person walking away from the radar.

close to the radar), higher for larger values of the range at about 2–3m distance, and lower again at high range near 4m (as the person was slowing down, having almost reached the end of the room). A different approach, based on short time Fourier transform (STFT) can also be applied to obtain a time-Doppler (TD) pattern that characterizes the velocities of different moving body parts over time. STFT performs several FFTs on the data using shorter, overlapped time windows so that each FFT produces a column of the final TD pattern over time. The key parameters of this operation are the duration of the short FFT window, the overlap factor, and the type of window chosen (e.g., Hamming, Hann, and so on), as they significantly influence how the final TD patterns appear. The absolute value of the squared result is typically known as the "spectrogram," and an example is shown in Figure 2.2. As expected, all Doppler contributions are negative, and a sinusoidal pattern can be seen due to the oscillation of the human torso while walking with superimposed peaks for the additional back and forth oscillating movements of the arms. This is a typical pattern for human gait, as widely shown in a great deal of published work that exploited these signatures for characterization and classification of movements (Amin 2018).

As TD patterns make it possible to characterize the smaller movements of individual body parts over time (as opposed to the Range-Doppler that tends to capture bulk movements), the concept of "*micro-Doppler*" signature is often used to describe this phenomenon (Chen, Tahmoush, and Miceli 2018). Micro-Doppler refers to the small frequency modulations on top of the bulk Doppler shift of a moving object; these are caused by the movements of body parts, specifically for human beings the swinging of limbs and torso, small head motion, hand gestures. Micro-Doppler signatures, often represented by their TD patterns or spectrograms, have been widely used in the context of assisted living and health care to identify daily activity patterns, detect fall events, and assess the gait quality of monitored people (Cippitelli et al. 2017; Wang et al. 2014; Amin et al. 2016; Le Kernec et al. 2019). More recently, additional domains of information have also been exploited reconsidering the information that the "range domain" may include, especially thanks to the development of mm-wave radar systems providing larger bandwidth and therefore finer range resolution, as discussed in section 2.2.1. Figure 2.3 shows a non-exhaustive list of possible radar data domains, adding to the conventional three formats shown in Figure 2.2, Cadence Velocity Diagrams, cepstrogram and Mel-Frequency Cepstral coefficients, range-Doppler surfaces, and radar cubes (Le Kernec et al. 2019). The use of raw radar data, in their complex I and Q components, has also been considered. A comprehensive explanation of all these radar data domains and the signal processing to obtain them goes beyond the scope of this chapter, but interested readers are referred to the review paper (Le Kernec et al. 2019) and relevant references within for additional details. Figure 2.3 also lists possible classification algorithms based on shallow and deep learning techniques, such as convolutional neural networks (CNN), stacked and convolutional auto encoders (SAE, CAE), long-short term memory (LSTM) networks, and stacked gated recurrent units (SGRU), as well as multidimensional principal component analysis (MPCA) as a tool to compress the relevant information from the data. The following section will describe some fundamental principles of machine and deep learning techniques applied to radar data.

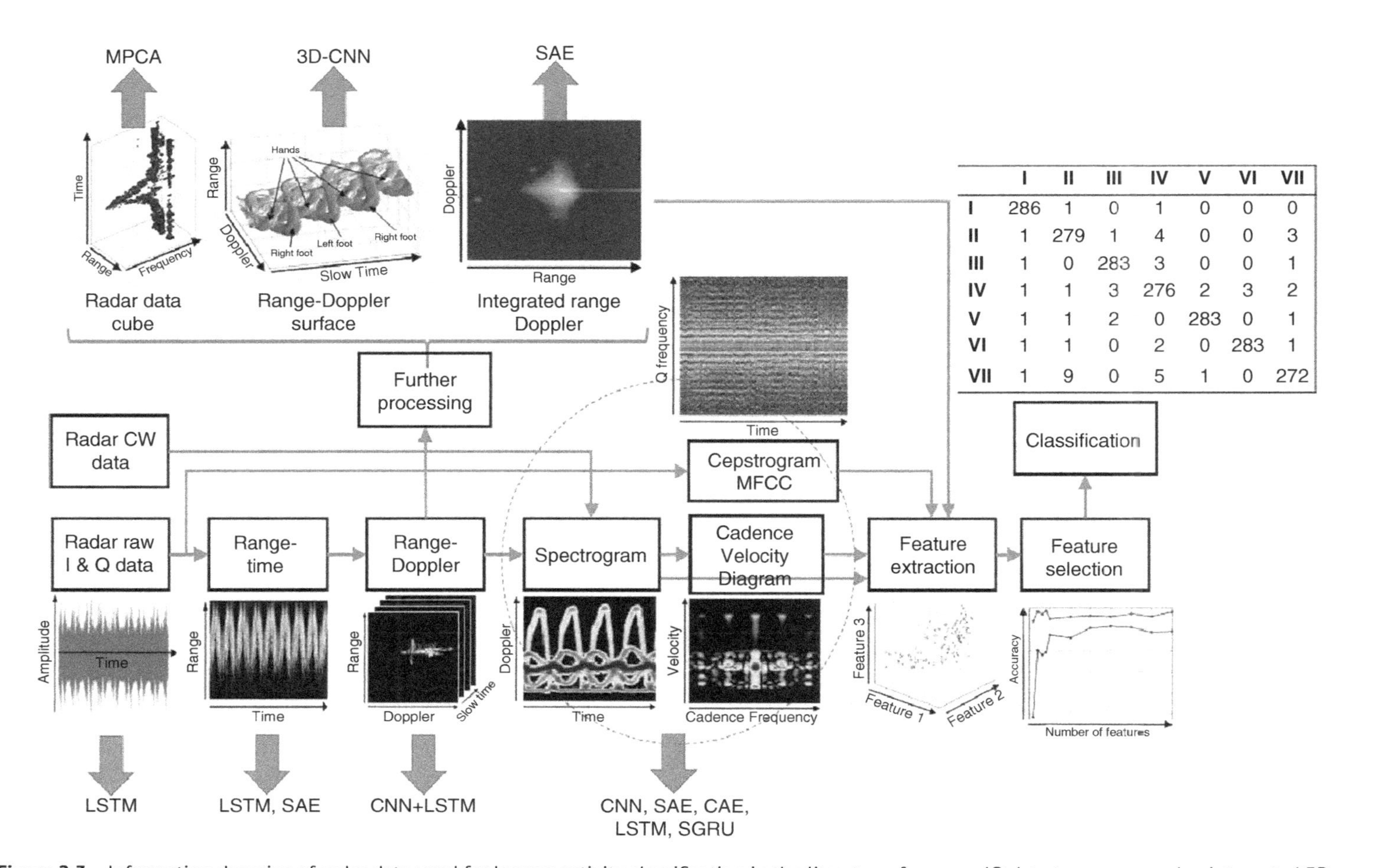

	I	II	III	IV	V	VI	VII
I	286	1	0	1	0	0	0
II	1	279	1	4	0	0	3
III	1	0	283	3	0	0	1
IV	1	1	3	276	2	3	2
V	1	1	2	0	283	0	1
VI	1	1	0	2	0	283	1
VII	1	9	0	5	1	0	272

Figure 2.3 Information domains of radar data used for human activity classification in the literature, from raw IQ data to more complex integrated 3D cubes and range-Doppler surfaces. J. Le Kernec et al., © 2019, IEEE.

2.2.3 Principles of Machine Learning Applied to Radar Data

Statistical machine learning and deep learning (LeCun, Bengio, and Hinton 2015) have witnessed terrific success and impact in recent years and have revolutionized numerous different fields in science and engineering, RF and radar sensing included. While a thorough discussion of the mathematical details of all these techniques goes beyond the scope of this chapter – there are indeed books exclusively dedicated to the topic such as (Hastie, Tibshirani, and Friedman 2017; Goodfellow, Bengio, and Courville 2016)– this section aims to provide a synthetic but yet comprehensive background on how these techniques can be used when applied to radar data, with particular emphasis on healthcare-related applications. Journal papers such as (Le Kernec et al. 2019; Gürbüz and Amin 2019) (provide additional details, written by radar engineers with the perspective of using statistical and deep learning tools with cognition and awareness.

The typical processing aims to use machine learning to teach algorithms how to automatically interpret and classify the patterns and signatures extracted from the radar data, essentially the images or the sequences of radar data in the formats described in Figures 2.2 and 2.3. This can be configured as a supervised learning problem, where an algorithm trained using existing labeled data is asked to recognize and classify a set of unknown testing data (Jain, Duin, and Mao 2000). A common framework for both statistical learning (SL) and deep learning (DL) is represented in Figure 2.4, with the three steps of data collection and pre-processing feature extraction, and actual classification.

For radar data, the pre-processing stage includes the formatting of the data in the most suitable domain among those discussed in the previous section. The "optimal" data representation is still an outstanding research question, as the aim is to maximize classification and recognition performances, but also to ensure that the method is portable across different scenarios and subjects. The general objective of this stage is to maximize the differences between different classes of interest (for example, different human activities) while minimizing inter-class variations (for example, the inevitable differences in the ways different people will perform the same activity, hence generating different radar signatures).

The feature extraction stage aims to compress the pre-processed data into a lower-dimensional representation, in order to capture into few quantitative numerical parameters the most significant information for the classification problem at hand. These numerical parameters are called features, and at the outcome of the feature extraction stage each original data sample is mapped into a point in an N-dimensional feature space, where N is the number of extracted features as illustrated in Figure 2.3. The objective of this stage is to find features that enable simple decision boundaries between classes in this N-dimensional space, where these boundaries are learned at the training stage based on the probability distributions of available samples for each class of interest.

The classification stage then implements a form of cost or loss function minimization of a new unseen test sample x belonging to a possible class c_i, accounting for boundaries of separation between classes. The way this function is calculated and minimized depends on the specific classification algorithm chosen, but performances can be assessed in terms of general metrics such as sensitivity, specificity, accuracy, and F1-score, among others. These metrics are listed for reference in Figure 2.4 for the case of a simple binary classification. Conventional classification algorithms include Bayes' theory (NB), K-nearest neighbors

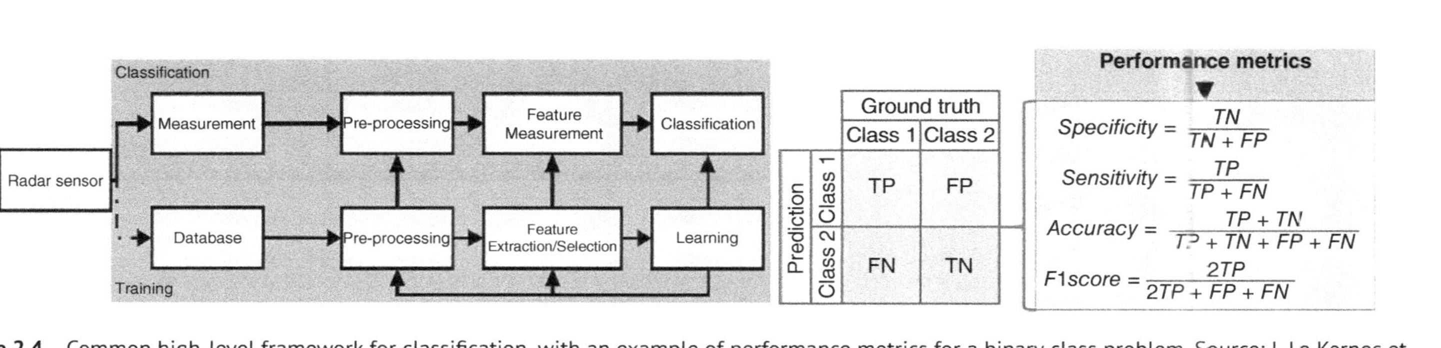

Figure 2.4 Common high-level framework for classification, with an example of performance metrics for a binary class problem. Source: J. Le Kernec et al., © 2019, IEEE.

(KNN), support vector machines (SVM), classification trees (CT), all described in (Hastie, Tibshirani, and Friedman 2017).

Assessing which features and how many are optimal for a given classification task is not a trivial problem, as too many features can provide redundant and conflicting information, thus reducing the final classification performance, also increasing the computational load for their extraction and processing. Hence, feature selection techniques can be used, such as "wrapper" methods that test different combinations with a specific classifier to select the best-performing features, "filter" methods that rank them based on inter-/intra-class information metrics (for example, entropy, Fisher score, T-tests), and techniques based on Principal component analysis (PCA). A vast number of features have been proposed in the literature: physical features aiming to reconstruct some kinematic parameters of the movements observed (Kim and Ling 2009), features based on the centroid and bandwidth of the micro-Doppler spectrograms (Fioranelli, Ritchie, and Griffiths 2016) or more complex yet effective moments inspired from image processing techniques (Persico et al. 2016), features based on Cepstral coefficients and their modifications to fit the specific characteristics of human radar signatures (Gürbüz et al. 2015; Erol, Amin, and Gürbüz 2018), among others.

In any case, all these can be considered "handcrafted" features: the initial choice as to what features to extract and how to implement this process was conventionally left to the expertise of the radar engineer and operator. With deep learning, some aspects of the pre-processing stage and the whole process of feature extraction and selection are automatically performed by the neural networks, bypassing the need for too many inputs and parameters set by radar operators and implicitly performing feature extraction and selection. Neural networks based on deep learning still require a large number of parameters, the "hyper-parameters," to be set, such as learning and dropout rate, number of epochs, number of hidden units and layers, and choice of activation functions, but existing work from the computer science community can support this process. The usage of deep learning has shown to provide a step-change in classification tasks based on radar data, including in the context of fall detection and more in general analysis of daily activities (Gürbüz and Amin 2019). Popular deep learning techniques for radar data include:

- Convolutional neural networks (CNNs) that exploit convolutional filtering layers to capture the salient features from radar data interpreted as an image, that is, a matrix of pixels, typically a spectrogram as Time-Doppler representation;
- Auto-encoders (AE) that are made of an encoder plus decoder stage, where the former maps the input data (typically an image) in a compressed space using a non-linear function, and the latter attempts to reconstruct the input from there. By removing the decoder part, the encoder can be used as a feature extractor block to be fine-tuned at the training stage for a specific classification problem. AEs can include convolutional layers at the encoder and their inverse at the decoder to implement convolutional AEs, CAEs;
- Recurrent neural networks, such as long-short term memory (LSTM) cells or gated recurrent units (GRUs) inspired from work in audio and speech processing, which interpret the radar data as a temporal sequence of pulses or time bins in a spectrogram, rather than as an image, 2D matrix of pixels.

A common characteristic of deep neural networks is the need for a very large number of data for proper training, given the many hyper-parameters to tune. If this is not a particular problem for the image and audio processing communities, thanks to the availability of large, shared datasets usable as common benchmark, it is an issue for radar data. The challenges of collecting and labeling such data, more complex than taking a picture or recording an audio file, and the lack of a sizeable, shared dataset of radar data are currently limiting factors (Fioranelli et al. 2019). Mitigating approaches include the use of feature extraction techniques that can operate well with fewer samples such as the AEs, the usage of transfer learning, and the generation of synthetic radar data to complement the small set of available experimental data. Transfer learning exploits neural networks that are already pre-trained on a different domain, for example, optical images, and performs fine-tuning of parameters at the training stage with the small set of radar data available; this has been shown to be a viable approach compared with random initialization of the parameters in deep networks (Gürbüz and Amin 2019). Synthetic or simulated data to augment the set of experimental radar data can be implemented using model-based approaches, such as those that model the human body as a series of discrete moving joints and generate the corresponding radar signatures (Ishak et al. 2019), or data-driven approaches including generative adversarial networks (GANs) that can generate credible but different enough replicas from small sets of real data (Garcia Doherty et al. 2019). All the aforementioned approaches have been shown to be interesting and with significant potential, but their improvement and validation is still an outstanding research challenge.

2.2.4 Complementary Approaches: Passive Radar and Channel State Information Sensing

The discussion in section 2.2 so far has assumed the usage of a conventional radar, where there are a transmitter and a receiver unit, that could be co-located and share part of the electronics components, but in any case the system would work by transmitting and receiving specific radar waveforms. Recently, passive radar systems have been proposed for several applications, including healthcare-related usages for indoor and short-range scenarios (Griffiths and Baker 2017; Kuschel, Cristallini, and Olsen 2019). These do not need a dedicated transmitter but exploit a transmitter of opportunity, which can be another radar system (an approach often referred to as hitchhiking radar), or a communication system (for example, Wi-Fi, Digital Video Broadcasting – Terrestrial (DVB–T), Frequency Modulation radio (FM), Global Navigation Satellite System (GNSS) satellites signals). The presence and distance of objects are typically determined by performing a cross-correlation operation to calculate the delay between the signal directly obtained from the emitter of opportunity (reference channel) and the echoes received after bouncing off potential objects in the area under-test (surveillance channel). If the emitter of opportunity operates with digital modulations, then the originally transmitted signal can be directly reconstructed from the decoded received signal using knowledge of the standard, so that only one channel, the surveillance one, is needed, as opposed to non-digital schemes where both reference and surveillance channels are needed.

The advantage of passive radar is its simpler hardware without the transmitter's relatively high-power electronics, and no need to allocate dedicated RF spectrum for the transmission. The disadvantage is no control on the characteristics of the transmitted waveforms, which are primarily not optimized to provide good radar performance in terms of range-Doppler ambiguity and target detection, but for communication metrics. Examples of passive radar techniques for health care based on Wi-Fi signal include the detection of vital signs such as respiration, and daily activities recognition (Chen et al. 2016a; Chen et al. 2016b). The signal-processing approach for feature extraction and classification is very similar for active and passive radar systems, with the techniques described in section 2.2.3 and some of the data formats in section 2.2.2 (e.g., spectrograms) still applicable to passive radar data.

Another approach involves the characterization and analysis of the wireless channel state information (WCI or CSI), as the presence of subjects moving in an area of interest generates perturbations and changes in the condition of the wireless propagation channel, such as propagation delays, amplitude attenuations, and Doppler shifts induction. Each movement, from macro-motions such as walking or sitting/standing to micro-motions such as those of the chest and abdomen while breathing, induces specific changes to the different channels of the Wi-Fi communication system. Hence these can be considered as "signatures" for the classification of such movements (Shah et al. 2018). An advantage of this approach is that relatively low-cost devices are needed, such as network interface cards embedded into conventional computers or modified Wi-Fi routers, without the need for additional radar-based transmitters and/or receivers. The potential disadvantage lies in the lack of a dedicated system and waveforms designed for monitoring purposes and the challenge in establishing robust classification algorithms based on WCI data.

In initial works, the typical approach used the received signal strength indicator (RSSI) from the different wireless channels to discriminate activities, but this approach based on strength (amplitude) of the signal is susceptible to noise and provides only coarse resolution. More recently, complex amplitude and phase information from a group of OFDM (orthogonal frequency division multiplexing) sub-carriers of the Wi-Fi signal has been used in this context, with a necessary pre-processing step of calibrating the phase information to make it usable. Exploiting complex information over a number of diverse sub-carriers provides additional information on amplitude, phase, and frequency shifts than simple RSSI, enabling more granular monitoring of the activities and movements performed by subjects in the area of interest (Shah et al. 2018). As seen for radar data, machine learning techniques that aim to extract significant features and apply classification algorithms can also be used on the WCI amplitude and phase data.

2.3 Radar Technology in Use for Health Care

This section presents a few use cases of radar technology demonstrated for healthcare applications, aiming to present some of the most recent results from leading researchers in this domain. The radar used were off-the-shelf systems produced by Ancortek (C-band FMCW radar) and Xethru (X-band pulsed ultra-wideband radar).

2.3.1 Activities Recognition and Fall Detection

Fall detection has been one of the first applications of contactless radar sensing in the context of assisted living (Cippitelli et al. 2017), as falls frequently become catastrophic events with aging. Fall events lead to immediate physical injuries such as cuts, abrasions, and fractures of bones, as well as to a psychological impact leading to the fear of falling again and, in general, to reduced confidence and diminished level of physical activity. Furthermore, a reduction in life expectancy is also reported for those who experience a long-lie period after the fall, that is, an involuntary rest on the ground for an hour or longer. Fall detection systems need to provide good detection rates, that is, not failing to detect any actual falls, and very low false alarm rates, that is, not being confused by activities that may generate similar signatures, for example, those where the subject sits down or bends down, with the body presenting an acceleration towards the ground. With radar systems, this may be a challenge, as all these movements are translated into similar Doppler shifts. Figure 2.5 shows an example of six Doppler–time patterns (spectrograms) for six different activities

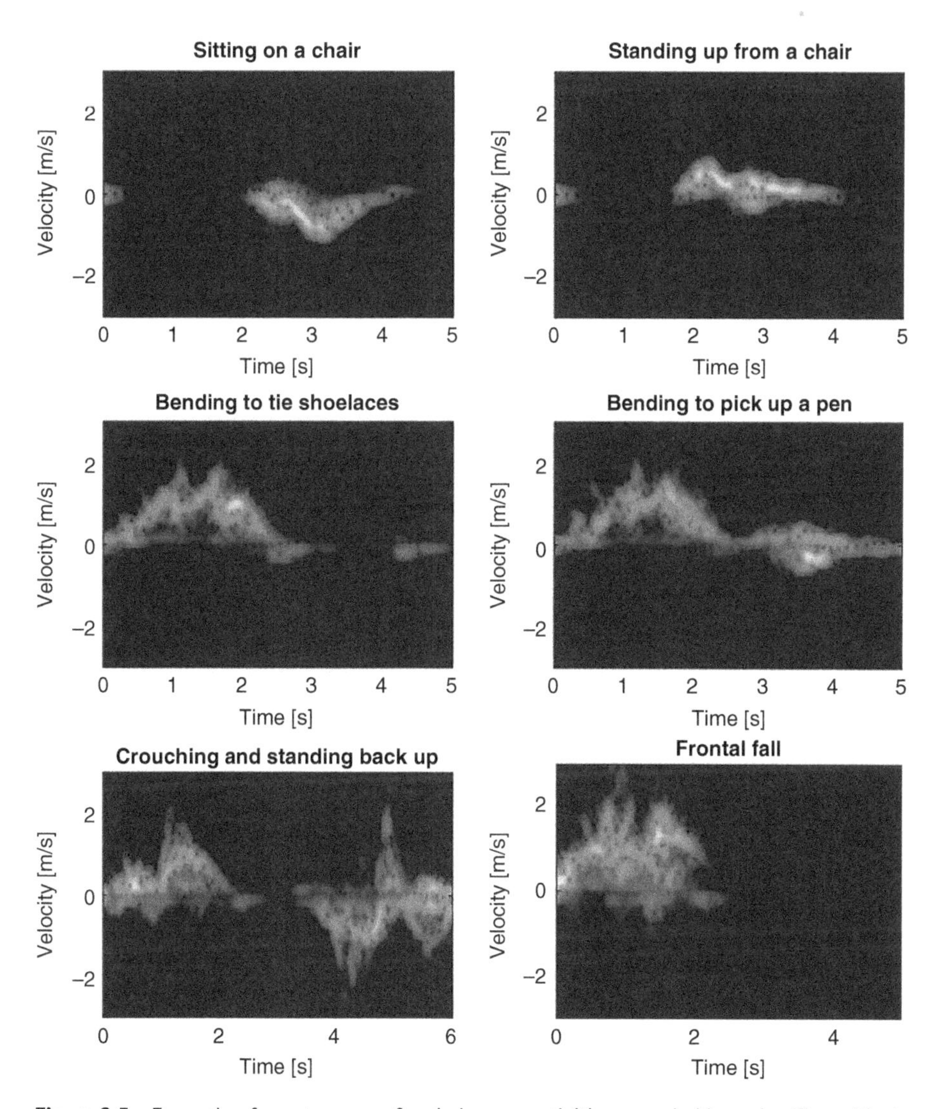

Figure 2.5 Example of spectrograms for six human activities recorded by radar. The subject (a) sitting on a chair, (b) standing up from a chair, (c) bending to tie shoelaces, (d) bending to pick up a pen, (e) crouching and standing back up, and (f) falling frontally – Source: F. Fioranelli, J. Le Kernec, and S.A. Shah © 2019, IEEE.

performed by the same subject while facing the radar, as recorded by a system operating in C-band (5.8 GHz) at the University of Glasgow. The six activities include sitting on a chair (Fig. 2.5a), standing up from a chair (Fig. 2.5b), bending to tie shoelaces (Fig. 2.5c), bending to pick up a pen (Fig. 2.5d), crouching and standing back up (Fig. 2.5e), and falling frontally after tripping (Fig. 2.5f). The activities were performed in parallel to the antenna beam, that is, with an aspect angle of approximately zero degrees with respect to the radar line of sight to maximize the recorded Doppler. The antennas used in this case were off-the-shelf Yagi designed for 5.8 GHz Wi-Fi communications, deployed at a height of approximately 1m from the ground and used with vertical polarization.

Activities that imply movement toward the radar, such as bending down and falling, all have positive velocity in their patterns, and vice versa (for example, the sitting on a chair scenario, where the subject, in this case, sat down and leaned a bit back on the chair, therefore generating a negative Doppler shift). The simulated fall (Fig. 5f) presents a strong positive signature similar to the activities of bending down (Figs. 5c and 5d) and crouching (Fig. 5e, initial part), hence the challenge of establishing an algorithm for automatic classification of these cases, accounting for all the variability of the different activities and different people.

To cope with this challenge, different techniques have been proposed and are summarized in (Cippitelli et al. 2017; Amin et al. 2016) where the features (that is, the relevant information for classification) and the classifiers used are described. These include a number of signal-processing approaches to format the radar data in a suitable domain (STFT for spectrograms, but also fractional implementation of the FFT, wavelet transformations, and more complex time-frequency transformations such as the EMBD, extended modified beta distribution), features ranging from handcrafted physical parameters extracted from spectrograms to more abstract quantities obtained from subspace decompositions of the radar data (e.g., Singular value decomposition) and Mel-Frequency Cepstral Coefficients inspired from audio processing, and classifiers, ranging from relatively simple KNN and SVM, to the use of deep learning, with CNNs interpreting the radar spectrograms as input images for the network and typically outperforming the simpler classifiers. In terms of the radar systems used for fall detection, initial studies used bench equipment such as VNAs (vector network analyzers) operating as CW radar to collect data for algorithmic validation, as well as commercial radar systems, the majority of which was operating at the 5.8 GHz and 24 GHz (Industrial, Scientific, and Medical bands to avoid potential issues with transmission into licensed parts of the electromagnetic spectrum) (Cippitelli et al. 2017).

Supported by the enhanced classification capabilities provided by deep learning approaches, the fall detection issue has recently been approached as a more general challenge of recognizing the different activities of daily living performed by people. From the healthcare perspective, monitoring patterns of activities can provide useful information about the general well-being of people, indicating how physically active they are (sedentary/active) and if they perform activities associated with normal cognitive capabilities (activities of daily living), such as food preparation and intake, sleep, and personal hygiene. Knowing the daily pattern of activities can also help detect anomalies and at times subtle changes that the person themselves may struggle to identify or report to their general practitioner (GP), but that can be related to worsening health conditions (for instance, a classic example can be an increased use of the toilet for urinary infection, or the time spent sleeping for instance, because of depression, cognitive/physical decline).

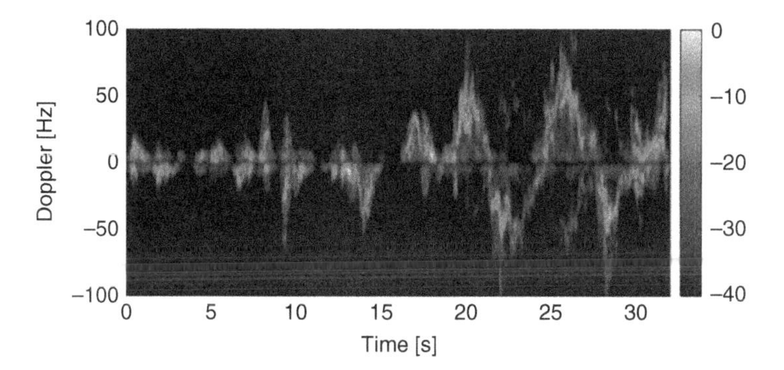

Figure 2.6 Spectrogram for a continuous sequence of six activities performed by a subject: drinking a glass of water while standing, picking up an object from the floor, sitting on a chair, standing back up, walking back and forth, and falling frontally – Source: F. Fioranelli, J. Le Kernec, and S.A. Shah © 2019, IEEE.

One of the recent challenges when moving from detecting just one activity (falls) to recognizing the different daily activities is the processing of a continuous stream of data, where the transitions between different activities can happen at random times, and there may be periods of inactivity between them. Figure 2.6 shows the spectrogram of an example of six activities performed one after the other, namely drinking a glass of water while standing, picking up an object from the floor, sitting on a chair, standing back up, walking back and forth, and falling frontally. The first four activities are fundamentally performed without much bulk movement of the body, so the Doppler signature is concentrated around the 0-Hz value, whereas the positive and negative contributions due to walking back and forth are visible between 20 and 30 s, with the final strong positive Doppler signature due to the fall at the end of the recording. While the processing of these streams of continuous activities is still an outstanding research challenge (Li et al. 2019), significant results have been achieved to classify different activities recorded as "snapshots," independent samples of data where only activity is performed such as those shown in the spectrograms of Figure 2.6.

One of the seminal papers in the domain of recognition of activities using radar is the work by Kim and Ling (2009), where six handcrafted features were extracted from spectrograms of seven human activities to achieve classification accuracy in the region of 91%. The significance of this paper was to present a then-novel application of radar sensing, as well as touching on important aspects such as feature selection, the importance of the aspect angle of the subjects' movements with respect to the radar, and the possibility of working with sequences of activities. These aspects were and are still thoroughly investigated in research. In terms of applications of deep learning methods to outperform manual feature extraction and selection, Seyfioglu, Ozbayoglu, and Gürbüz (2018) present a detailed analysis of three approaches based on AEs, CNN, and CAEs, compared with a more conventional approach of using SVM with a selected pool of 50 features. These different classification approaches were evaluated on a large set of 12 human activities performed by 11 subjects collected with a 4 GHz CW radar. CAEs were shown to provide better performance than AEs or CNN on their own, with accuracy of about 94%, whereas the conventional approach with the SVM could only yield results in the region of 77%. The work in (Li et al. 2019; Li et al. 2018) showed how radar information can be complemented by other sensors in an information-fusion framework, with particular emphasis on wearable sensors (e.g., accelerometers, magnetometers, and gyroscopes) that may be worn by some users without issues of compliance/acceptance. The results show an increase in accuracy of up to above 96% when information from both sensors is combined using feature or decision fusion, for the analysis of a set of 10 different activities including walking, sitting and standing, carrying objects, bending and crouching, and a simulated fall, among others.

2.3.2 Gait Monitoring

Another application of radar sensing in health care is the analysis of gait and locomotion parameters. Comprehensive gait analysis is typically performed, especially for older people, as part of a broader fall risk assessment, commonly done by trained personnel and clinicians in a highly specialized gait laboratory. While these tests in a specialized laboratory can provide very rich details, these assessments are expensive and uncomfortable for people taking part, as they have to travel to the location of the laboratory. Hence, this type of assessment is rather infrequent and may also generate inaccurate results, as people will behave and walk differently in an unusual laboratory environment compared to their homes (Wang et al., 2014). While in-home gait assessments are possible as part of routine primary care visits by GP or trained nurses, unbiased comparisons between different visits and objective records of the gait parameters over time are challenging. In this context, radar sensing can provide unobtrusive and continuous gait monitoring in private home environments, enabling early detection of changes that can increase risks of falling and show degradation in overall health conditions.

Typical applications in the literature have seen the use of spectrograms to identify the pattern of the legs' movements while the subjects are walking and to extract relevant parameters to measure the overall gait speed, any asymmetry or limping, the length of the stride, and the number of steps people can comfortably walk before needing to stop. Figure 2.7 shows a simple example of four different walking gaits performed by the same subject, namely normal stride, dragging one foot, short steps, and long stride; in all four cases, the subject was walking away from the radar, hence the initial net Doppler shift close to zero for the first couple of seconds, followed by negative Doppler signatures. As expected, an abnormal gait with short steps or dragged foot slows down the person, with more seconds necessary to cover the same distance and vice versa for the case of a longer stride. The number of steps can be counted by the number of negative peaks in the spectrograms, associated with the movement of individual legs and feet while walking; as expected, the case of shorter steps has many more peaks compared with the case of a long stride where fewer steps are needed to cover the same physical distance. It should be noted that these representations of the radar data show the Doppler shift, hence velocity, as a function of time. Therefore, the extraction of distance/location information such as the length associated with each step is not straightforward, but can be performed looking at the range-time representation of the data (as commented in section 2.2), provided that the radar has enough resolution.

Besides this empirical observation, research work has currently been undertaken to derive more precise gait parameters from radar signatures, such as periodicity of the gait, mean velocity and acceleration, stride length, and step time, as these can have clinical values in assessing the more general health condition of patients at risk. The work in

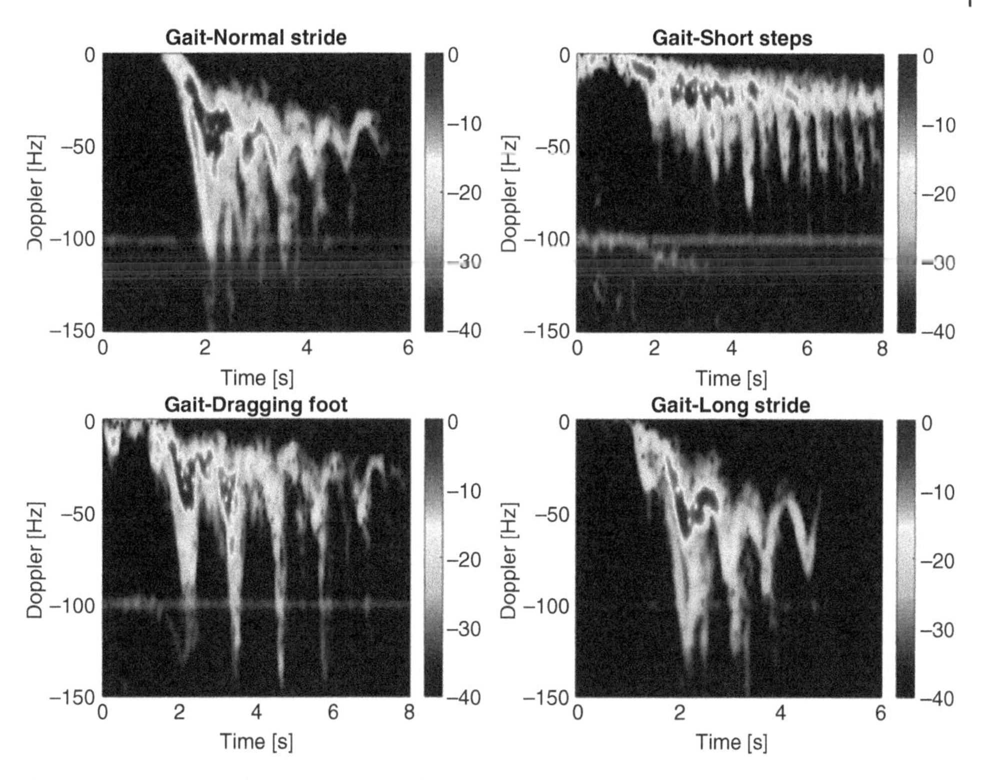

Figure 2.7 Example of spectrograms for four different walking gait performed by the same subject in the same scenario. Source: F. Fioranelli, J. Le Kernec, and S.A. Shah, © 2019, IEEE.

(Wang et al. 2014) was among the first to derive parameters such as walking speed and step time using Doppler radar sensing, demonstrating through the analysis of data from 13 participants very good statistical agreement with metrics generated by a Vicon motion capture system, which is typically used in specialized gait assessment laboratories. The recent work in Seifert, Amin, and Zoubir (2019) proposes a classification approach based on radar data to distinguish and classify five different types of gait for each person, including normal, pathological, and assisted walks with canes. Average accuracy values above 92% are demonstrated, as well as good generalization capabilities for the algorithm when data from an unknown individual are used for testing, that is, subjects whose data were not used to train the classification algorithm. An earlier piece of work in (Gürbüz et al. 2017) considered five different classes of mobility gait, namely normal unaided walking, walking with a limp, walking using a cane or tripod, walking with a walker, and moving using a wheelchair. This can be relevant, as people may need to transition from an unaided to an aided gait (insurgence of locomotion problems) or vice versa (during or just after recovery and rehabilitation stage, for example), or there may be different people living in the same environment (e.g., a nursing home) and their individual gait has to be identified first before extracting any gait parameter. While the classification results in terms of accuracy in distinguishing these five gaits were in the region of 80–85%, an interesting element of this paper is the simultaneous analysis of data from different radar systems at 5.8, 10, and

24 GHz, as well as an ultrasound SONAR system operating at 40 kHz. Another interesting paper is Saho et al. (2019), where the authors presented a study to correlate gait metrics extracted from radar micro-Doppler signatures with the results of standard tests to measure cognitive abilities of people. The significance of the work is that rather than monitoring cognitive capabilities through such time-consuming cognitive function tests, one could instead use the simple features extracted from the radar signatures: for example, mean and maximum velocities of the legs and of the body, and the standard deviation of the leg velocity. The work was validated across data with 74 participants and showed statistically significant correlation, which is encouraging for further studies in this area to link cognition assessment and not just physical conditions assessment to the radar signatures.

2.3.3 Vital Signs and Sleep Monitoring

Besides macro-movements such as those discussed in the previous two sections for activities and walking gait, radar sensing has been demonstrated for detection and monitoring of vital signs, including respiration and breathing disorders, heartbeat, sleep stages through small movements of the body, and possible neurological conditions through monitoring of tremors. The sensing principle is the same as for macro-activities, that is, the detection of movements through the changes in the phase of the received radar signals and the related Doppler shifts; however, the movements are much smaller in this case, hence the need of finer precision and superior hardware capabilities to characterize these signals. As an example, Figure 2.8 shows the spectrogram for the respiration of a human subject sitting on a chair at about 60 cm from the radar. The subject simulated different respiration rates: 10 usual inhaling/exhaling cycles, followed by holding their breath for several seconds, then a deep exhaling, and finally a few cycles of accelerated breathing. The distinction between

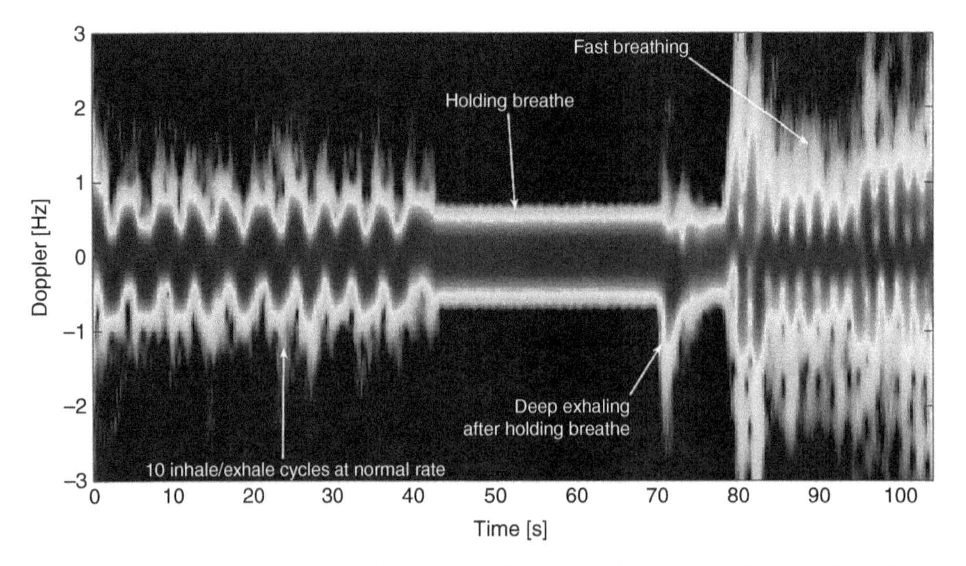

Figure 2.8 Example of spectrogram for a person sitting on a chair at approximately 60 cm from the radar and simulating different respiration rates. *Source*: Fioranelli, Le Kernec, and Shah (2019) with slight modifications.

the different respiration rates and regimes is quite clear in the spectrogram, and the periodicity and regularity of respiratory movements can be extracted from these data. The number of normal inhale/exhale cycles (10) was directly counted by the subject during the measurement as a form of ground truth prior to holding the breath. The number of fast breathing cycles were not counted in this case, as the goal was just to show empirical differences in the spectrogram between the two regimes in terms of Doppler frequencies excited and the distribution of energy.

The advantage of radar sensing for these applications is to free the subjects from the discomfort of probes attached to them, making it possible to extend the monitoring period and to reduce or avoid biases due to stress or physical constraints caused by contact-based monitoring equipment. Some outstanding research challenges include investigating the robustness of the monitoring algorithm for vital signs with respect to the distance of the subject, the orientation and position of the body (for example, aspect angle between the subject and the radar line of sight, as well as positions such as sitting rather than lying, in cases where the subject is lying on their side, or in prone/supine positions), the presence of layers of clothes worn by the subject or objects such as bed linens, blankets, or curtains.

There is extensive literature on the use of radar sensing for respiration and cardiac monitoring using radar, for example, (Li et al., 2017) on applications of short-range contactless sensing that include vital signs of animals (such as laboratory animals like rat or farmed animals such as dairy cows) besides human, or (Quaiyum et al. 2018) where an electromagnetic model is proposed and validated along with experimental radar data to enable generalization of the results and fine-tuning of the proposed radar signal processing algorithms. A comprehensive discussion on Doppler radar for physiological sensing is provided in this book (Boris-Lubecke et al. 2016) for the readers' reference, where an analysis of the hardware requirements and architectures for the radar systems in these applications are presented.

Beyond monitoring the presence or absence of respiration and the respiration rate, breathing disorders can also be investigated using radar sensing to provide additional information on health conditions for diagnosis and prognosis. These disorders are generally characterized as irregular breathing patterns and may be induced by a range of conditions, such as respiratory muscle weakness, stroke, heart-related diseases, metabolic imbalances, injuries to brain respiratory centers, or usage of certain medications (Zhao et al., 2018). As discussed earlier, typical devices for monitoring of breathing disorders have to be worn by the patients, causing discomfort and at times preventing long-term monitoring in home environments or during the night; hence, the attractiveness of using contactless radar sensing. The work in (Zhao et al. 2018) proposed a CW radar system for breathing disorders monitoring, consisting of a bespoke designed 2.4 GHz radar-sensing module and a recognition module to analyze the recorded radar waveforms and identify possible breathing disorders based on a multiclass SVM classifier. An interesting feature of this work was its validation with real patients in the Nanjing Chest Hospital in China, where a patient diagnosed with nocturnal breathing disorders was involved in the experiment. A good agreement between the automatic predictions provided by the proposed radar system and the manual labels of an expert clinician was noted, although some errors were recorded due to random body movements while sleeping. The removal of random body movements is indeed an important challenge for radar-based vital signs monitoring, as these may generate larger Doppler shifts and signal fluctuations compared to the small signals related to the vital signs (Boris-Lubecke et al. 2016).

Sleep stages monitoring is another growing application of radar sensing. There is an emerging body of literature connecting poor sleep quality with adverse health effects at physical and cognitive levels, hence the recent interest in monitoring sleep quality and the different stages in sleep to identify potential problems. Sleep medicine uses polysomnography (PSG) as the gold standard to characterize sleep stages, but this technique requires a specialized laboratory and trained sleep technicians, which is unfeasible for large-scale monitoring of many potential patients. Recently, radar sensing has been proposed as an alternative approach that promises to be simpler and easier to scale even in private home environments. The core idea is to extract features that are measurable from the radar data (such as respiration and heart rate, their mean and variance, the presence of body movements, and other physiological signals) and correlate them back to the different sleep stages such as deep sleep and REM (rapid eye movement) phase (Hong et al. 2018). A simple CW Doppler radar was used to validate this idea, where the baseband I/Q radar signals produced by the hardware module are analyzed by a "sleep stage estimation block" to extract features and perform classification, providing at the final output labels to distinguish wakefulness, light sleep, deep sleep, and REM phase sleep. The system was validated in a six-hour long experiment to compare the outputs from the proposed radar with the labels from a polysomnography machine considered as the ground-truth (Hong et al. 2018). The proposed approach used 11 features derived from body movements and respiration and heart rate, processed by a subspace KNN classifier; labels were provided every 60 seconds. The results showed accuracy rates in the region of 76–89% for the different sleep stages, a promising result that can be improved with additional tests and research on more advanced feature extraction and classification techniques.

2.4 Conclusion and Outstanding Challenges

This chapter has described some of the most recent results of radar RF sensing in the context of health care, for applications including recognition of activity patterns and fall detection, abnormal gait analysis, and monitoring of vital signs and sleep quality. Although conventionally associated with large-scale, defense-related uses to monitor ships or aircraft, these examples show how radar sensing is now a promising technology for a number of short-range, civilian applications.

However, some outstanding research challenges remain to be addressed and are briefly discussed below:Addressing the potential problem of multi-occupancy, that is, the simultaneous presence of multiple subjects or people with their pets whose radar signatures can overlap in range and/or velocity over time, with the need to separate them to extract individual healthcare-related information.

- Incorporating the additional information made available by recent mm-wave compact radar systems, which typically can provide large bandwidths of several GHz, hence very fine spatial resolution, and access to multiple transmitting/receiving channels to implement beam forming and direction of arrival techniques. This can help address the previous point on multi-occupancy, as well as make it possible to resolve contributions to the radar echo from different body parts directly in the range domain.

- Implementing the whole process of radar signal processing plus information extraction and classification in real time and in a constrained computational environment, for example, embedded platforms, leveraging on "sparse" or "on-edge" computation. This is necessary for translation of the emerging research approaches into deployable and cost-effective products and services, superseding the offline implementation on dedicated desktop machines, which are typically seen in research work.
- Adapting established classification techniques from the analysis of individual, separated activities into the characterization of a more realistic, continuous, and uninterrupted sequence of actions. This can draw on the inherent working principles of radar sensing, that is, the continuous transmitting and receiving of a sequence of radar waveforms, rather than "forcing" the classification algorithm to interpret the data as a "snapshot" optical image. In this approach, the challenge is not only the classification of individual activities, but also the detection of transitions between them, periods of absence of activity, and anomalies with respect to the normal routine. Furthermore, when dealing with continuous sequences of activities, different classification approaches may be required for different activities, for example, for macro-movements (walking, sitting and standing, moving objects, food preparation) as opposed to smaller motion (recognition of gestures) or the monitoring of the very small motion associated with vital signs (respiration for example). An intelligent system should be able to switch to the relevant classification approach while monitoring the flow of activities as the person performs them. A further challenge when dealing with long sequences of many activities is the related labeling, that is, how to generate enough labeled data to support classification algorithms based on supervised learning frameworks like those described in this chapter. To address this, unsupervised learning approaches that can cluster and organize the data without explicit labels may be very suitable.
- Providing a comprehensive and robust validation of any proposed technique by including realistic environments and procedures beyond controlled laboratory spaces, and involving actual end-users ensuring diversity of age, gender, and physical conditions. This is to develop algorithms and systems capable of dealing with such complex yet realistic diversity. For this aspect, there may be two potentially conflicting requirements. On the one hand, one could train a system based on as much data from as many people as possible to increase its robustness and generalization capabilities, so that the system can have a good knowledge of what radar signatures for a "normal human" look like. On the other hand, the system should be over-fitted and fully personalized to monitor a specific individual in their environment, in order to check any anomaly with respect to the normal routine of that person. To some extent, the system should replicate the operations of a good GP, who knows the acceptable average values for a given health-related parameter across the whole population, but also the individual circumstances and personalized conditions of their patients.

2.5 Future Trends

With the advent of 5G and beyond, our living environments will become more connected than ever with the Internet of Things (IoT), Wi-Fi, and indoor deployment of millimeter wave networks to support upcoming telecommunication standards. Radar will become a sensor in a suite of sensors, as one among the other IoT devices in a smart home. 5G is foreseen as a key enabler for indoor sensing networks with device-to-device communication and mobile edge computing, which enable the capacity and the flexibility to manage the recording and manipulation of radar data. One of the technical challenges to unlock the healthcare-related sensing applications discussed in this chapter lies in the trade-off between the amount of radar data exchanged among the different parts of the system, and the amount of pre-processing required before each transmission (Le Kernec et al. 2019). This will have an impact on embedded platforms such as radar processing units as well as other sensing modalities in terms of real-time processing, energy consumption, computational power, and communication capabilities. In Latf et al. (2017a) and Latf et al. (2017b), the authors describe all the change opportunities in the healthcare industry that mobile health (mHealth) can enable for applications such as remote surgery, e-health, ambient patient monitoring, supported by artificial intelligence, cloud/cognitive computing, commoditization of data from crowdsourcing, and next-generation communication. This step change is actually in looking at prevention by detecting precursor symptoms to a condition rather than just reacting/detecting problems as they occur, as shown with technology-integrated health management for dementia in the example in (Rostill et al. 2018). This paradigm shift is vital for the healthcare system and its sustainability in the future. There is a shift from specialized care centers towards caring for patients within their home environment.

This evolution in sensing also has to be reflected in the development of future monitoring systems for in-home patient monitoring. The consequences from the perspective of radar and RF sensing are two-fold:

1. Paradigm change in radar sensing
2. Multimodal sensing

2.5.1 Paradigm Change in Radar Sensing

Pervasive deployment of millimeter wave communication equipment indoors, Wi-Fi, and miniaturized radar front-ends with multiple-input multiple output opens new challenges for radar sensing. The classical approach in radar is monostatic, with one transmitter and one receiver co-located in the same device. New systems in millimeter wave offer miniaturized systems with multiple transmitters and multiple receivers, which allows enhancing spatial resolution beyond what the instantaneous bandwidth can offer. Despite radar units becoming more accessible, driven by the automotive industry, prototypes remain costly and usually do not fully match the application requirements. Wi-Fi and 5G access points are very good transmitters of opportunity that can be exploited for passive radar sensing. This will multiply the possible radar configurations and their diversity dramatically.

One way of testing various radar configurations in different situations would consist of simulating the environment, activities and radar system in order to test its fitness for

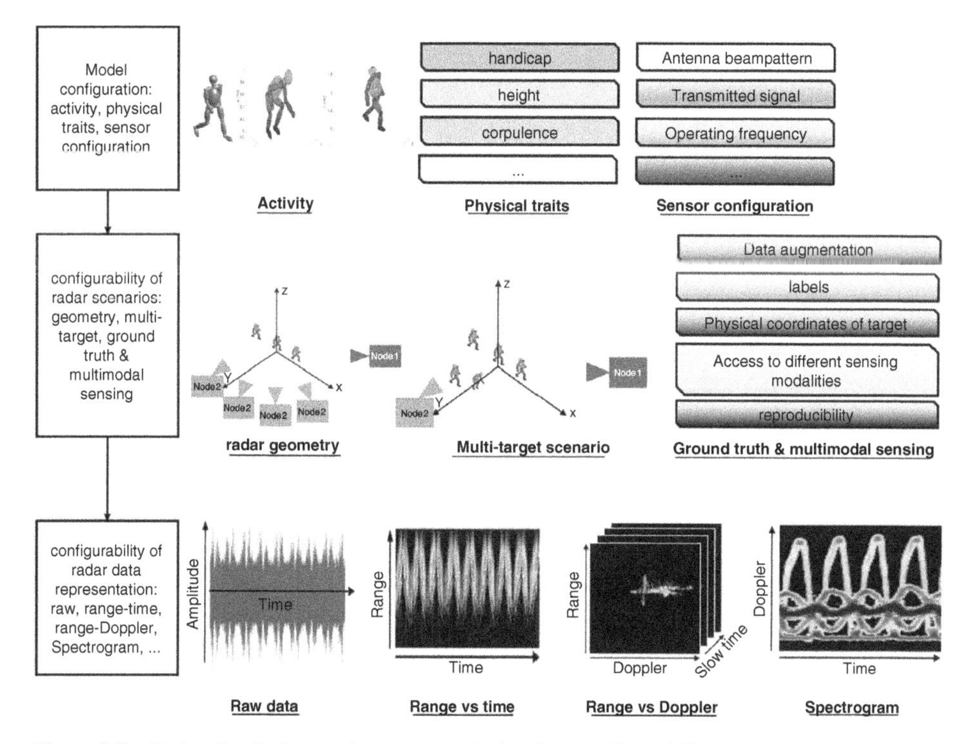

Figure 2.9 Radar simulation environment to design for paradigm shift.

purpose as illustrated in Figure 2.9. Unlike experimental trials, simulation requires fewer resources in terms of equipment, manpower, and cost.

The model is configurable in terms of activity, human model, and sensor configuration. There are a number of kinematic models and motion-capture databases such as those described in Chen, Tahmoush, and Miceli (2018); Ishak et al. (2019); and Lin et al. (2018) that can be used to ensure the fidelity of the simulation to physical phenomena. These can be configured to take into account height, corpulence, and disabilities or constrained movements, for example, to vary physical traits, as they may have an influence on classification accuracy. The radar sensor configuration can be defined within the model, taking into account the operating frequency, bandwidth, the transmitted signal (chirp, OFDM, 5G, etc.), the antenna beam patterns, and more.

Additionally, the scenarios for the simulations can be configured to take into account:

- The geometry (monostatic, multistatic, passive, interferometric, aspect angle of the target with respect to the radar),
- The multi-target scenarios to take into account multi-occupancy in a home to include pets, spouses, multi-generational homes, and visitors

For all the scenarios, the ground truth is available at all instants in time, which may not be possible for live experiments. By ground truth is meant the exact position from the sensor(s), trajectory, labels of the activities, and the possibility to get further sensing modalities by deriving acceleration at any point of the human model, for example. Other sensing modalities than radar could be integrated in simulation. Furthermore, the simulator can be used

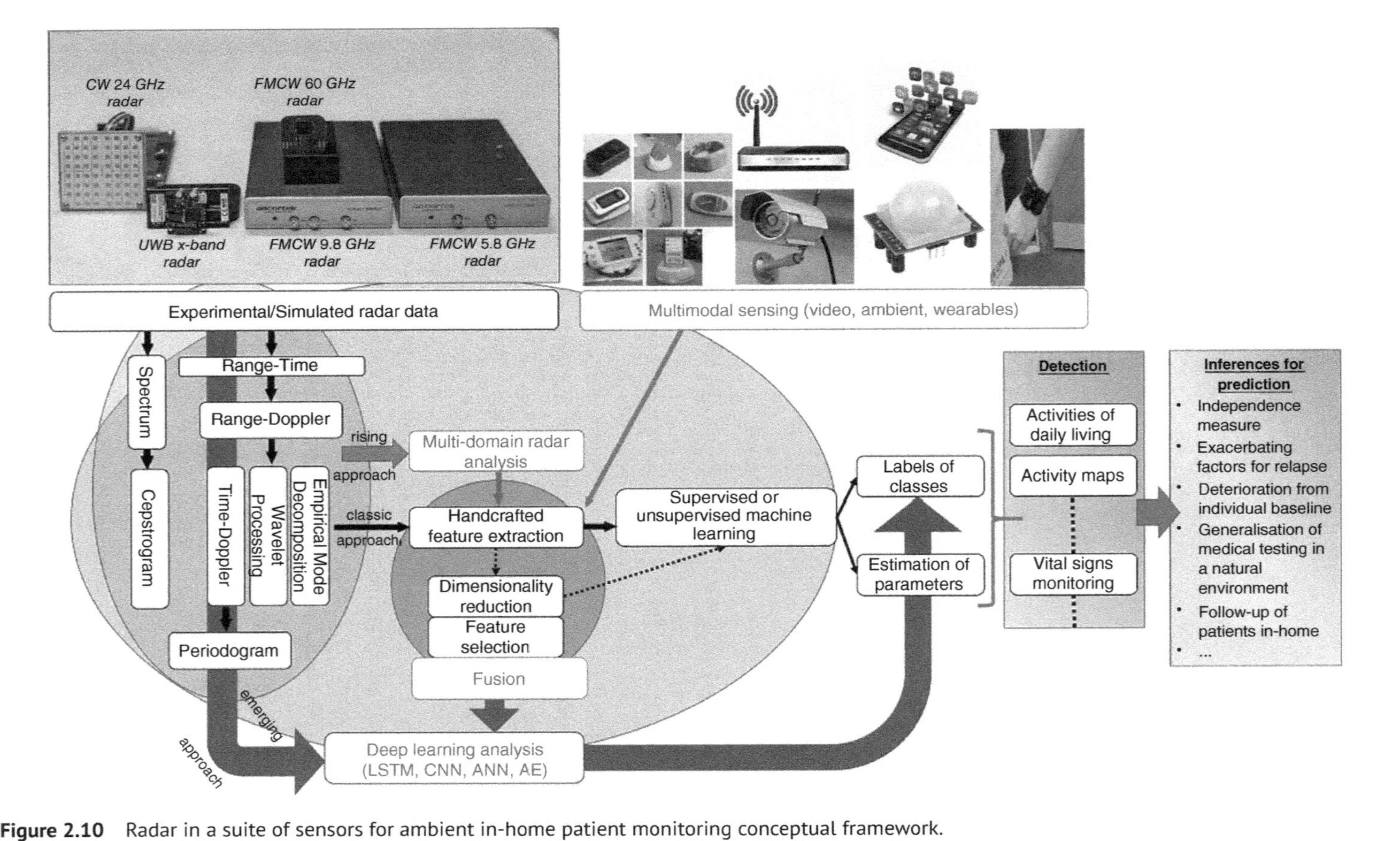

Figure 2.10 Radar in a suite of sensors for ambient in-home patient monitoring conceptual framework.

to augment databases for limited datasets to increase performance by modifying the pace of the target and the physical traits of the human model, as well as synthetic sample generation (Erol, Gürbüz, and Amin 2019; Chawla et al. 2002). Be aware that samples generated in this way should be checked for kinematic fidelity. This means that the synthetic samples should respect the phenomenology and that samples generated with the wrong features artificially may affect negatively the performances of the classifier.

Finally, the radar data representation domain from which features are extracted for classification can also be configured from raw data, range time, range-Doppler, spectrogram, and other representations, to capture more relevant features from these diverse domains of radar data (Le Kernec et al. 2019).

Note that simulations will never replace experimental measurements for validation of actual performances, but this allows dimensioning a system based on available information from limited resources. This is especially important when exploring future radar systems to develop new concepts especially when the technology does not currently exist, such as THz radar or 5G and beyond access points with associated receivers for passive operation.

2.5.2 Multimodal Sensing

Radar will become "a sensor in a suite of sensors" within the ambient space of a patient or, more in general, a vulnerable subject at home. This means that other ambient sensors will be monitoring the environment of the patient, looking, for example, at thermal comfort, connected appliances with connected TVs, cooker sensors (on/off), communication devices (e.g., Wi-Fi, 5G access points), wearable sensors (smartphone, connected watches), implants for an ambient IoT environment specific to each home. As these devices will be providing different information at different data rates, their information will need to be properly fused in order to increase the accuracy of prediction models.

Machine learning algorithms will need to become adaptive to take into account sensors joining and leaving the ambient IoT sensing of the patient dynamically, and the need to consider diverse processing and features extraction/selection on the same sensor as the person monitored performs different activities. Furthermore, the adaptive capability will need to deal with sensors that stop working, sensors working in a synchronous or asynchronous manner, and sensors being moved or wearables being worn or not. This changes the way algorithms are designed to predict a class of actions or estimate vital signs in order to infer information for prevention (Fig. 2.10).

References

Adib, F., H. Mao, Z. Kabelac, D. Katabi, and R.C. Miller. (2015) "Smart Homes That Monitor Breathing and Heart Rate," in *Proceedings of the 33rd annual ACM conference on human factors in computing systems*, p.837–846.

Amin, M. G. (ed.) (2018) *Radar for indoor monitoring*. Boca Raton, FL: CRC Press.

Amin, M.G., Y.D. Zhang, F. Ahmad, and K.C.D.D. Ho (2016) *Radar signal processing for elderly fall detection: the future for in-home monitoring*, 33(2), p.71–80.

Amin, M.G., Y.D. Zhang, F. Ahmad, and K.C.D. Ho (2016) "Radar signal processing for elderly fall detection: the future for in-home monitoring," *IEEE Signal Processing Magazine*, 33(2), p.71–80.

Aykan, H. (2012) "*Report to Congress: Aging services technology study*," U.S. Dep. Heal. Hum. Serv. Off. Disabil. Aging Long-Term Care Policy.

Barnett, K., S.W. Mercer, M. Norbury, G. Watt, S. Wyke, and B. Guthrie. (2012) "Epidemiology of multimorbidity and implications for health care, research, and medical education: a cross-sectional study," *Lancet*, 380(9836), pp. 37–43, 2012.

Boris-Lubecke, O., V.M. Lubecke, A. D. Droitcour, B.K. Park, and A. Singh (eds.). (2016) *Doppler radar physiological sensing*. Hoboken, NJ: Wiley.

Buxi, J, D.. Redouté, and M.R. Yuce. (2017) "Blood pressure estimation using pulse transit time from bioimpedance and continuous wave radar," *IEEE Trans. Biomed. Eng.*, 64(4), p.917–927.

Chaccour, K., R. Darazi, A. H. El Hassani, and E. Andrès. (2017) "From fall detection to fall prevention: a generic classification of fall-related systems," *IEEE Sensors Journal*, 17(3),p.812–822, 2017.

Chawla, N.V., K.W. Bowyer, L.O. Hall, and W.P. Kegelmeyer. (2002) "SMOTE: synthetic minority over-sampling technique," *Journal of Artificial Intelligence Research*, 16, p.321–357

Chen, C., Y. Han, Y. Chen, H.-Q. Lai, F. Zhang, B. Wang, and K.J.R. Liu. (2018) "TR-BREATH: time-reversal breathing rate estimation and detection," *IEEE Trans. Biomed. Eng.*, 65(3), p.489–501.

Chen, Q., K. Chetty, K. Woodbridge, and B. Tan (2016b) "Signs of life detection using wireless passive radar," 2016 IEEE Radar Conference (RadarConf), p. 1–5.

Chen, Q., B. Tan, K. Chetty, and K. Woodbridge. (2016a) "Activity recognition based on micro-Doppler signature with in-home Wi-Fi," 2016 IEEE 18th International Conference on e-Health Networking, Applications and Services (Healthcom), p. 1–6.

Chen, V.C., D. Tahmoush, and W.J. Miceli (eds.) (2018) *Radar micro-Doppler signatures: processing and applications*. London: Institution of Engineering and Technology.

Cippitelli, E., F. Fioranelli, E. Gambi, and S. Spinsante. (2017) "Radar and RGB-depth sensors for fall detection: a review," *IEEE Sensors Journal*, 17(12), p.3585–3604.

Debes, C., A. Merentitis, S. Sukhanov, M. Niessen, N. Frangiadakis, and A. Bauer. (2016) "Monitoring activities of daily living in smart homes: understanding human behavior," *IEEE Signal Processing Magazine*, 33, no. 2, p.81–94.

Dong, S., Y. Zhang, C. Ma, and C. Zhu. (2019) "Doppler cardiogram: a remote detection of human heart activities," *IEEE Trans. Microw. Theory Tech.*, p.1–10.

Ebrahim, M. P., F. Heydari, K. Walker, K. Joe, J. Redoute, and M. R. Yuce. (2019) "Systolic blood pressure estimation using wearable radar and photoplethysmogram signals," in 2019 IEEE International Conference on Systems, Man and Cybernetics (SMC), p. 878–3882.

Erol, B., M.G. Amin, and S.Z. Gürbüz. (2018) "Automatic data-driven frequency-warped cepstral feature design for micro-Doppler classification," *IEEE Trans. Aerosp. Electron. Syst.*, 54(4), p.1724–1738.

Erol, B., S.Z. Gürbüz, and M.G. Amin. (2019) "GAN-based synthetic radar micro-Doppler augmentations for improved human activity recognition," IEEE Radar Conference, April 22–26, p. 1-6, Boston, MA, USA.

Fioranelli, F., J. Le Kernec, and S.A. Shah. (2019) "Radar for health care: recognizing human activities and monitoring vital signs," *IEEE Potentials*, vol. 38, no. 4, 2019.

Fioranelli, F., M. Ritchie, and H. Griffiths. (2016) "Centroid features for classification of armed/unarmed multiple personnel using multistatic human micro-Doppler," *IET Radar, Sonar & Navigation*, 10(9), p.1702–1710.

Fioranelli, F., M. Ritchie, S.Z. Gürbüz, and H. Griffiths. (2017) "Feature diversity for optimized human micro-Doppler classification using multistatic radar," *IEEE Trans. Aerosp. Electron. Syst.*, 53(2), p.640–654.

Fioranelli, F., S.A. Shah, H. Li, A. Shrestha, S. Yang, and J. Le Kernec (2019) "Radar sensing for healthcare," *Electron. Lett.*, 55(19), p.1022–1024(2).

Garcia Doherty, H., L. Cifola, R. Harmanny, and F. Fioranelli (2019) "Unsupervised learning using generative adversarial networks on micro-Doppler spectrograms," European Radar Conference (EURAD), 2019.

Goodfellow, I., Y. Bengio, and A. Courville (2016) *Deep learning*. Cambridge, MA: MIT Press.

Griffiths, H., and C. Baker. (2017) *An introduction to passive radar*. London: Artech House.

Gürbüz, S.Z., and M.G. Amin. (2019) "Radar-based human-motion recognition with deep learning: promising applications for indoor monitoring," *IEEE Signal Process. Mag.*, 36(4), p.16–28.

Gürbüz, S.Z., C. Clemente, A. Balleri, and J.J. Soraghan. (2017) "Micro-Doppler-based in-home aided and unaided walking recognition with multiple radar and sonar systems," *IET Radar, Sonar Navig.*, 11(1), p.107–115.

Gürbüz, S.Z., B. Erol, B. Çağlıyan, B. Tekeli, B. Cağliyan, and B. Tekeli (2015) "Operational assessment and adaptive selection of micro-Doppler features," IET Radar, Sonar & Navigation, 9(9), 2015, p. 1196–1204.

Hastie, T., R. Tibshirani, and J. Friedman (2017) *The elements of statistical learning - data mining, inference, and prediction*, 2nd ed. Springer.

Hong, H., L. Zhang, C. Gu, Y. Li, G. Zhou, and X. Zhu. (2018) "Noncontact sleep stage estimation using a CW Doppler radar," *IEEE Journal on Emerging and Selected Topics in Circuits and Systems*, 8(2), p.260–270.

Ishak, K., N. Appenrodt, J. Dickmann, and C. Waldschmidt. (2019) "Advanced radar micro-Doppler simulation environment for human motion applications," 2019 IEEE Radar Conference (RadarConf), p. 1–6.

Jain, A.K., R.P.W. Duin, and J. Mao. (2000). "Statistical pattern recognition: a review," *IEEE Trans. Pattern Anal. Mach. Intell.*, 22(1), p.4–37.

Kuwahara, M., E. Yavari, and O. Boric-Lubecke. (2019) "Non-invasive, continuous, pulse pressure monitoring method," in 2019 41st annual international conference of the IEEE Engineering in Medicine and Biology Society (EMBC), p. 6574–6577.

Kim, Y., and H. Ling. (2009) "Human activity classification based on micro-doppler signatures using a support vector machine," *IEEE Transactions on Geoscience and Remote Sensing*, 47(5), p.1328–1337.

Kuschel, H., D. Cristallini, and K.E. Olsen (2019) "Tutorial: passive radar tutorial," *IEEE Aerosp. Electron. Syst. Mag.*, 34(2), p.2–19.

Latif, S., J. Qadir, S. Farooq, and M.A. Imran. (2017a) "How 5GWireless (and Concomitant Technologies) Will Revolutionize Healthcare?," *Future Internet*, 9(4), 93.

Latif, S., R. Rana, J. Qadir, A. Ali, M.A. Imran, and M.S. Younis. (2017b) "Mobile health in the developing world: review of literature and lessons from a case study," *IEEE Access*, 5, p.11540–11556.

Le Kernec, J., F. Fioranelli , C. Ding, H. Zhao, L. Sun, H. Hong, J. Lorandel, and O. Romain. (2019) "Radar signal processing for sensing in assisted living: the challenges associated with real-time implementation of emerging algorithms," *IEEE Signal Process. Mag.*, 36(4).

LeCun, Y., Y. Bengio, and G. Hinton (2015) "Deep learning," *Nature*, 521(7553), p.436–444.

Lee, D., K. Smith, C. Csech, and G. Shaker, "Glucose concentration estimation using electromagnetic waves," in 2018 18th International Symposium on Antenna Technology and Applied Electromagnetics (ANTEM), p. 1–4.

Li, C., Z. Peng, T.-Y. Huang, T. Fan, F.-K. Wang, T.-S. Horng, J.-M. Munoz-Ferraras, R. Gomez-Garcia, L. Ran, and J. Lin. (2017) "A review on recent progress of portable short-range noncontact microwave radar systems," *IEEE Trans. Microw. Theory Tech.*, 65(5), p.1692–1706.

Li, H., A. Shrestha, H. Heidari, J.L. Kernec, and F. Fioranelli (2018) "A multisensory approach for remote health monitoring of older people," *IEEE J. Electromagn. RF Microwaves Med. Biol.*, 2(2).

Li, H., A., Shrestha, H. Heidari, J. Le Kernec, and F. Fioranelli (2019a) "Activities recognition and fall detection in continuous data streams using radar sensor," IEEE MTT-S 2019 International Microwave Biomedical Conference, IMBioC 2019 - Proceedings.

Li, H., A. Shrestha, H. Heidari, J. Le Kernec, and F. Fioranelli (2019b) "Magnetic and radar sensing for multimodal remote health monitoring," *IEEE Sens. J.*, 19(20), p.8979–8989.

Lin, Y., J. Le Kernec, S. Yang, F. Fioranelli, O. Romain, and Z. Zhao. (2018) "Human activity classification with radar: optimization and noise robustness with iterative convolutional neural networks followed with random forests," *IEEE Sensors Journal*, 18(23), p.9669–9681.

Ohata, T., K. Ishibashi, and G. Sun. (2019) "Non-contact blood pressure measurement scheme using doppler radar," in 2019 41st Annual International Conference of the IEEE Engineering in Medicine and Biology Society (EMBC), p. 778–781.

Omer, A. E., G. Shaker, S. Safavi-Naeini, K. Murray, and R. Hughson (2018) "Glucose Levels Detection Using mm-Wave Radar," *IEEE Sensors Lett.*, 2(3), p.1–4.

Persico, A.R., C. Clemente, L. Pallotta, A. De Maio, and J. Soraghan. (2016) "Micro-Doppler classification of ballistic threats using Krawtchouk moments," 2016 IEEE Radar Conference (RadarConf), 2016, p. 1–6.

Quaiyum, F., N. Tran, T. Phan, P. Theilmann A.E. Fathy, and O. Kilic. (2018) "Electromagnetic modeling of vital sign detection and human motion sensing validated by noncontact radar measurements," *IEEE J. Electromagn. RF Microwaves Med. Biol.*, 2(1), p.40–47.

Rijken, M., A. Hujala, E. van Ginneken, M.G. Melchiorre, P. Groenewegen, and F. Schellevis. (2018) "Managing multimorbidity: Profiles of integrated care approaches targeting people with multiple chronic conditions in Europe," *Health Policy (New. York)*, 122(1), p.44–52.

Rostill, H., R. Nilforooshan, A. Morgan, P. Barnaghi, E. Ream, and T. Chrysanthaki. (2018) "Technology integrated health management for dementia," *British Journal of Community Nursing* 23(10), p.502–508, 2018.

Saho, K., K. Uemura, K. Sugano, and M. Matsumoto. (2019) "Using micro-Doppler radar to measure gait features associated with cognitive functions in elderly adults," *IEEE Access*, 7, p.24122–24131.

Sakamoto, T., and K. Yamashita. (2019) "Noncontact measurement of autonomic nervous system activities based on heart rate variability using ultra-wideband array radar," *IEEE J. Electromagn. RF Microwaves Med. Biol.*, p. 1.

Seifert, A., M.G. Amin, and A.M. Zoubir. (2019) "Toward unobtrusive in-home gait analysis based on radar micro-Doppler signatures," *IEEE Trans. Biomed. Eng.*, 66(9) p.2629–2640.

Seyfioglu, M.S., A.M. Ozbayoglu, and S.Z. Gürbüz. (2018) "Deep convolutional autoencoder for radar-based classification of similar aided and unaided human activities," *IEEE Trans. Aerosp. Electron. Syst.*, 54(4), p.1709–1723.

Shah, S.A., D. Fan, A. Ren, N. Zhao, X. Yang, and S.A.K. Tanoli. (2018) "Seizure episodes detection via smart medical sensing system," *J. Ambient Intell. Humaniz. Comput.*, Nov.

Skolnik, M. (2008) *Radar handbook*, third edition. New York: McGraw-Hill Education.

Storey, A. (2018) "Living longer: how our population is changing and why it matters," *Off. Natl. Stat.*

Stove, A.G. (1992) "Linear FMCW radar techniques," *IEE Proc. F – Radar Signal Process.*, 139(5), p.343–350.

Wang, F., M. Skubic, M. Rantz, and P.E. Cuddihy. (2014) "Quantitative gait measurement with pulse-doppler radar for passive in-home gait assessment," *IEEE Trans. Biomed. Eng.*, 61(9), p.2434–2443.

Wu, S., T. Sakamoto, K. Oishi, T. Sato, K. Inoue, T. Fukuda, K. Mizutani, and H. Sakai (2019) "Person-specific heart rate estimation with ultra-wideband radar using convolutional neural networks," *IEEE Access*, 7, p.168484–168494.

Zhao, H., H. Hong, D. Miao, Y. Li, H. Zhang, Y. Zhang, C. Li, and X. Zhu. (2018) "A Noncontact breathing disorder recognition system using 2.4-GHz Digital-IF Doppler radar," *IEEE J. Biomed. Heal. Informatics*, 23(1), p.208–217.

3

Pervasive Sensing: Macro to Nanoscale

Qammer H. Abbasi[1,], Hasan T. Abbas[2], Muhammad Ali Imran[1] and Akram Alomainy[1]*

[1]*James Watt School of Engineering, University of Glasgow, Glasgow, United Kingdom*
[2]*School of Electronic and Electrical Engineering, Queen Mary University of London, London E1 4NS, United Kingdom*

Technological breakthroughs in the fields of nano-fabrication have enabled us to realize miniaturized communication systems. Nano-scale pervasive sensing is a vision through which vital physiological data of a patient are captured with the help of self-powered nano-sensors that are positioned on human skin. The collected data can then be used for a variety of diagnostic purposes. This chapter deals with the challenges involved in the deployment of these nano-scale sensing networks. Specifically, we discuss the terahertz frequency electromagnetic wave propagation on human skin with the help of which nano-scale communication takes places.

Technological breakthroughs in the fields of nano-fabrication have enabled us to realize miniaturized communication systems. Nano-scale pervasive sensing is a vision through which vital physiological data of a patient are captured with the help of self-powered nano-sensors that are positioned on human skin. The collected data can then be used for a variety of diagnostic purposes. This chapter deals with the challenges involved in the deployment of these nano-scale sensing networks. Specifically, we discuss the terahertz frequency electromagnetic wave propagation on human skin with the help of which nano-scale communication takes places.

3.1 Introduction

The advent of wearable electronics such as smartwatches and fitness trackers has made body-centric communication increasingly important and mainstream. A body-centric wireless network (BCWN) usually indicates a network formed by the nodes, spreading over a human subject in the topology of star or multi-hop, which communicates wirelessly with each other. It mainly consists of sensors and actuators (Latré et al., 2004), which can be divided into three domains depending on the location of the sensors: on-body communication, off-body communication, and in-body (in-vivo) communication (Akyildiz et al., 2008b). These networks can be used in various applications ranging from health care to

*Corresponding Author: Qammer.Abbasi@glasgow.ac.uk

Engineering and Technology for Healthcare, First Edition.
Edited by Muhammad Ali Imran, Rami Ghannam and Qammer H. Abbasi.

sports performance monitoring. Therefore, it enables constant monitoring of physiological data and persistent access to patients regardless of their whereabouts or activity. Moreover, it substantially reduces the cost associated with an examination involving a clinical visit. BCWN systems are particularly useful in case of remote at-home assistance, smart nursing homes, clinical trials, and research augmentation. It is estimated that remote medical diagnostics have the potential to save up to $25 billion worldwide in annual healthcare costs by reducing hospitalizations and extending independent living for seniors (Akyildiz et al., 2008b). The promising future of these technologies makes the study of antennas, radio signals, and communication systems for BCWNs an extremely attractive field of research, one that has been growing at a rapid pace for the past few years.

Nanotechnology has enabled the realization of nano-devices that gather vital information, and then communicate with external systems so that useful decisions can be made. Recently, there has been an effort to standardize the communication protocols used for nano-devices, and as a consequence nano-networks were proposed by the IEEE standardization group (P1906.1 - Recommended Practice for Nanoscale and Molecular Communication Framework) (IEEE, 2016). Nanotechnology, first envisioned by the Nobel laureate physicist Richard Feynman in his famous speech entitled "There's plenty of room at the bottom" in 1959 (Hey, 2018), has also helped the evolution of devices that range in scale from one to a few hundred nanometres. During the last few decades, emerging research areas such as nano-electronics, nano-mechanics and nano-photonics are allowing the development of novel nano-materials, nano-crystals, nano-tubes and nano-systems that promise to revolutionize many fields of science and engineering. Nanotechnology is a multi-disciplinary field with almost potential applications. First, in the biomedical domain, nano-particles such as dendrimers, carbon fullerenes (Buckyballs) and nano-shells are currently used to target specific tissues and organs (Kelly, 2013). Another area where nano-technology plays an important role is environmental science, where molecular and genomic tools are used to uncover the complexity of the induced defense signaling networks of plants (Binzoni et al., 2008). Furthermore, in industry molecular-scale filters and pores with well-defined sizes, shapes, and surface properties give engineers better functionality in molecular sieving (Berry et al., 2003a). The envisaged nano-machines are the most basic functional units able to perform very simple tasks at the nano-scale, including computing, data storage, sensing, actuation, and communication. In the electromagnetic (EM) domain, recent advances in nano-technology have allowed the implementation of devices such as graphene-based nano-sensors and lithium nano-batteries (Piro et al., 2015b). In the biological domain, living cells are a clear example of living nano-machines. Artificial nano-machines are expected to become a reality soon. A conceptual diagram of a nanomachine in the EM domain is shown in Figure 3.1.

Building a communication interface is critical to achieving the required functionality of nano-machines (or nano-devices). Nano-scale communication is referred to as the exchange of information at the nano-scale and, therefore, it is the basis of any wired/wireless interface of nano-devices in a nano-network. The type of communication interface is governed by the type of application, thus constraining the choice on the particular type of nano-communication. For the time being, several alternatives have been proposed, which can be divided into the following two categories (shown in Fig. 3.2); classic communication, which is mainly to downscale the existing communication paradigm, and biological communication, which uses molecules to encode, transmit, and receive information.

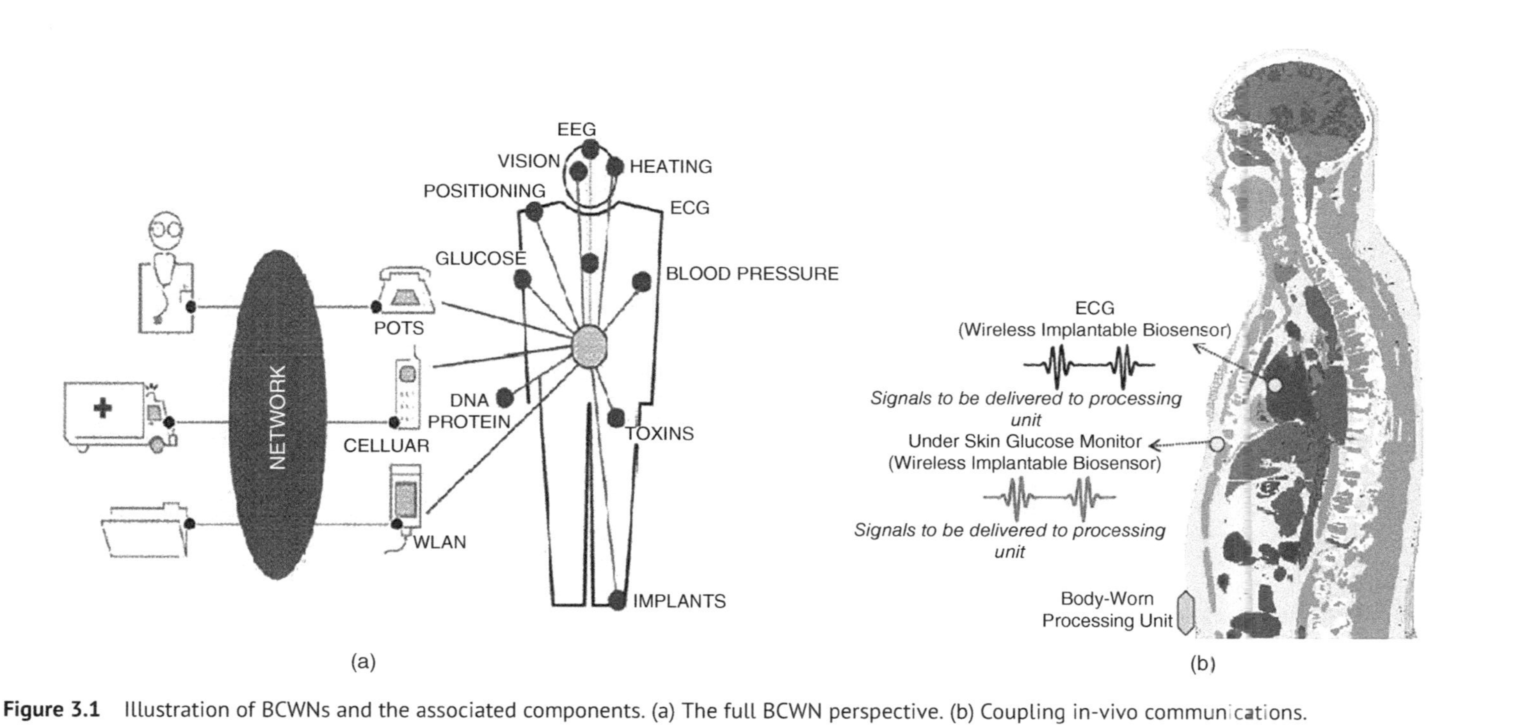

Figure 3.1 Illustration of BCWNs and the associated components. (a) The full BCWN perspective. (b) Coupling in-vivo communications.

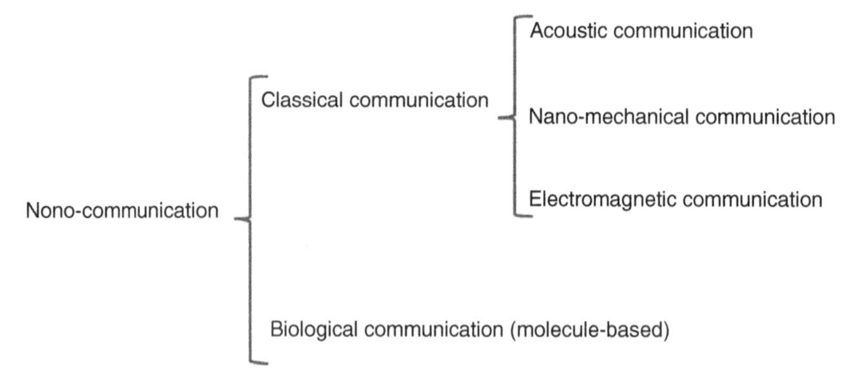

Figure 3.2 Categories of nano-scale communications.

Due to the minuscule size of the nano-machines, operating them is limited to their close nano-environment. To achieve meaningful functionality, a large number of nodes are required to be deployed in close proximity. These nano-machines will also need to control and coordinate their functions, leading to several research challenges in communication at the nano-scale. Nano-networks, the interconnection of nano-machines, provide means for cooperation and information sharing among them, allowing nano-machines to cover larger areas and fulfil more complex tasks. At the micro-scale, it is quite common to use EM waves, which can propagate with minimal losses using metal wires or through the air as media. Wiring a large quantity of nano-devices is not feasible; therefore, wireless solutions are more logical to use. Recently, researchers have attempted to investigate how nano-devices communicate using EM waves. Classical communication methods cannot be directly applied to nano-networks by just reducing the dimensions of basic elements of conventional networks because of the complexity and bulky size of the existing transceivers as well as their high power consumption. For instance, the influence of the quantum effect at nano-scale makes the application of the conventional network paradigm based on classical electronics impossible (Akyildiz et al., 2008b). An antenna made of graphene is not just a reduced-sized antenna of the traditional one; the resonant frequency of such nanostructures can be up to several orders lower than that of the non-carbon counterpart because several quantum phenomena can affect the propagation of EM waves in graphene.

In this chapter, we focus on the challenges faced when using EM waves for nano-scale communications. We also present a comprehensive overview of the framework required to develop channel models and determine the electronic properties of the physiological components such as the human skin that play a pivotal role in the realization of such networks. Moreover, we also discuss the characteristics of human skin and the frameworks behind the configuration of nano-networks.

3.2 The Anatomy of a Human Skin

The human skin layer can be divided into three major sections; epidermis, dermis, and fat with definitive thicknesses and functionality. However, the structure is far more complex

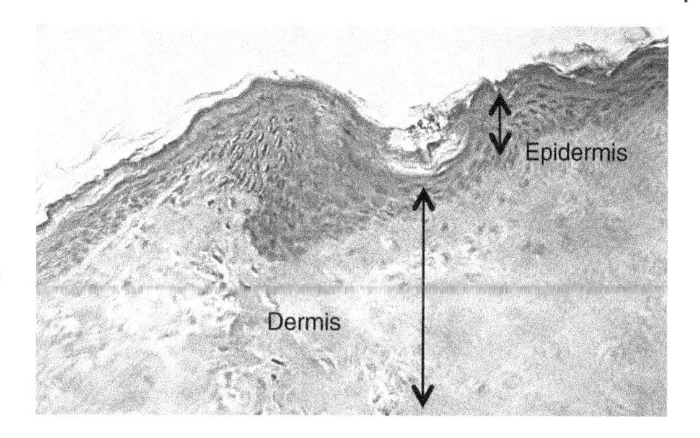

Figure 3.3 Microscopic image of a real human skin illustrating the two defined layers: epidermis and dermis. The sample sections were stained using Haematoxylin (purple/blue stain) and Eosin (red/pink stain), used for identifying nuclei and cytoplasm respectively. Stratum corneum traces were visible in the microscope; however, the thickness is not quantifiable.

and random. Skin in the human body is a protective layer, which can be considered a sensor of multiple parameters such as pressure, temperature, and so on. The epidermis is the thin outer layer (as shown in Fig. 3.3), which can be further divided into stratum corneum (SC), keratinocytes, and the basal membrane. The SC consists of mature keratinocytes, which contain the fibrous protein keratin. The layer just beneath it is keratinocytes containing live cells that mature and form the SC. The deepest layer of the epidermis is the basal membrane responsible for preparing new keratinocytes and replacing the old ones (Gawkrodger and Ardern-Jones, ; Blanpain and Fuchs, 2009). The thickest (3 mm) of all is the dermis (Fig. 3.3 lying just beneath the epidermis, which houses vital entities such as sweat glands, hair follicles (as shown in Figs. 3.4a and 3.4b), and nerve endings. This layer is rich in proteins such as elastin, fibrin, fibrinogen, and collagen. The collagen forms an extracellular matrix, acting as scaffolding for the dermis. The last layer of the skin is a network of collagen and fat cells, known as the subcutaneous layer.

For in-body networks, the skin should be investigated for three reasons: firstly, most of the in-body functioning or dis-functioning affects the skin (and water concentration). Secondly, skin is rich in biological entities and structures such as blood vessels, sweat ducts, capillaries, proteins, and so forth, and finally, it is easily accessible for measurement, making it a versatile subject of terahertz time-domain spectroscopy (THz-TDS).

3.3 Characterization of Human Tissue

Several studies on the applicability of THz communication for biomedical applications are performed in Berry et al. (2003b), Fitzgerald et al. (2003), Jornet and Akyildiz (2011), and Akyildiz and Jornet (2010a). The optical parameters of human tissues up to 2.5 THz have been empirically characterized in Berry et al. (2003b), while the possibility of applying EM waves in nano-networks is presented in Akyildiz and Jornet (2010a). Jornet el al. Jornet and Akyildiz (2011) studied the THz channel model in the air by varying water vapour concentration and also proposed a new physical-layer aware medium access control (MAC) protocol (Jornet et al., 2012). The characteristics of EM waves propagating inside a human body at THz frequencies have been studied in Yang et al. (2015), where it is shown that the path loss has a relation with the dielectric loss of human tissues. In Feldman et al. (2009), the electrical properties (conductivity and permittivity values) of each layer have been investigated depending on the water content of each layer.

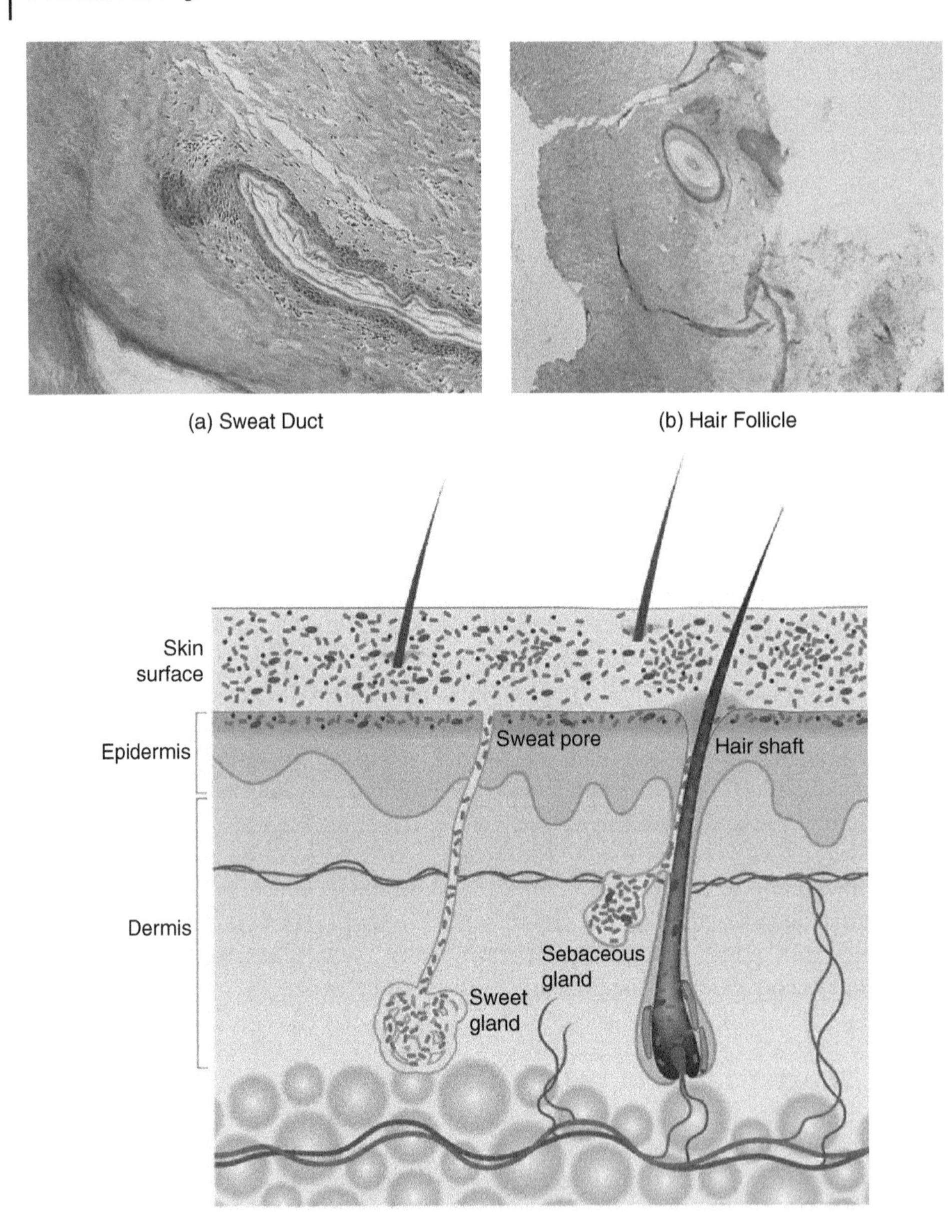

(c) Schematic of human skin with layered structure (Grice and Segre, 2011)

Figure 3.4 Cross-section of skin sample (a) Longitudinal section of the hair follicle attached to a sweat duct in the dermis layer, (b) Cross-section of oval-shaped hair follicle structure. The hair follicles are mainly present in the dermis layer acting as sensory receptors. (c) Illustration of different layers of human skin. Source: Elizabeth A Grice and Julia A Segre © 2011, Springer Nature.

In other studies, (Pickwell and Wallace, 2006; Chan et al., 2009; Smye et al., 2001), the spectral range of 0.1 to 10 THz is exploited for a wide range of biomedical applications, mainly to better understand the dynamics of water flow, security, dental, tissue imaging, and protein folding and unfolding. Strong absorption of radiation by water at THz frequencies is very sensitive to change in the water content of materials, and this can, therefore, be exploited for biomedical applications. Also, the THz region is considered as safe for such applications as compared to conventional microwave frequencies (Gallerano, 2004). Extensive studies have been performed on skin imaging at these frequencies; Pickwell et al. (2004) conducted an in-vivo terahertz imaging study from 0.1 to 1.4 THz to investigate the person-to-person variations of terahertz skin properties, and demonstrated how the thickness of the SC on the palm could be measured non-invasively. It has been reported in Hintzsche et al. (2012) that THz radiation with its non-ionizing and non-invasive innateness, when exposed to cells does not express any changes in DNA repair.

In the past, human skin dielectric properties have been studied extensively within the MHz to GHz range (Akyildiz and Jornet, 2010b; Akyildiz et al., 2008a), however, limited studies are presented on animal and human tissue dielectric properties in the THz domain using time-domain spectroscopy (TDS) (Yuce et al., 2007; Shnayder et al., 2005). In-body nano-networks and nano-communication are novel domains of the internet of nano-things with applications in health care. However, certainly one needs to revise the basic communication theory and establish its understanding at a nano and cellular level. Some studies have been presented on such nano-devices proposing them to be graphene-based (Jornet and Akyildiz, 2010). The transmission schemes have been investigated in the presence of lossy air medium and result in favor of THz communication offering very high physical transmission rates (1Tb/s) and transmission distance of the order of few tens of mm (Piro et al., 2015a). Therefore, it is imperative to understand the fundamental EM behaviour of the tissue samples while achieving repeatability in results.

In Gabriel et al. (1996), network arrangement is proposed to be comprised of nano-nodes, router, interface devices, and gateway. Tiny devices such as nodes and routers can be exploited from the naturally existing biological units at a molecular level to establish hybrid molecular-EM communication. Molecular communication is based on the concepts of diffusion and is a different study altogether.

To characterize the human skin in the THz band, we conducted an experimental study using the THz-TDS material characterization technique to extract the human skin material properties. The EM properties of the human skin distinctively affect the propagation features as can be seen by the path loss models, which are vital for channel modelling of in-body nano-networks. We need to determine two parameters, that is, the refractive index and the absorption coefficient to completely define the electrical properties of the skin, which includes the way EM waves propagate in the skin. Using the THz-TDS technique, the data are recorded in the time domain, which is converted to the frequency domain by taking its Fourier transform (FT). As an example, the measured time-domain signals for a sample mimicking the human skin, air and TPX are shown in Figure 3.5, and it can be seen that the signal is highly attenuated in the biological sample. The transmission through TPX is used as reference data with its main peak at 15.7 ps and an attenuated secondary peak at 27.6 ps. To extract the material properties, it is essential to have a time delay between the reference and sample data. Figure 3.5 shows that that the sample is shifted with respect to the TPX with its main peak at 20 ps and satellite peak at 31.9 ps. The oscillations and attenuation in the data are due to the presence of water vapors in the environment.

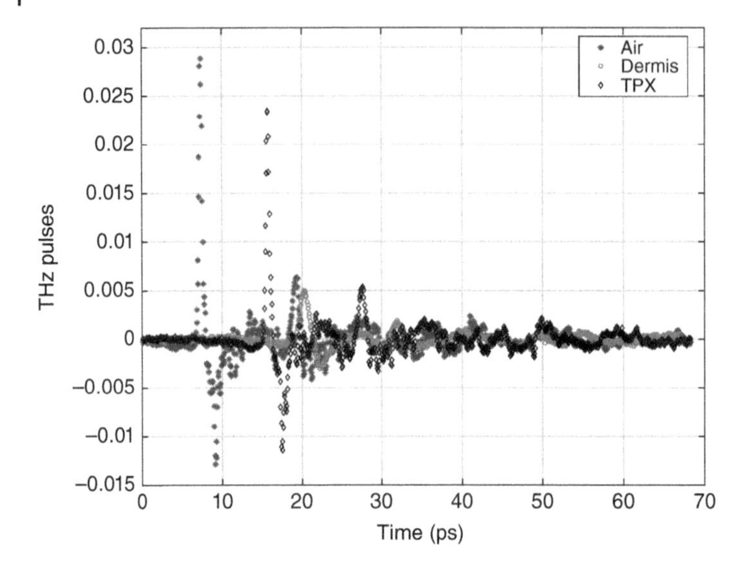

Figure 3.5 Measured THz pulses through air, TPX, and a human tissue sample. As expected the biological sample is highly attenuated, and the second peak is almost lost due to absorption with a maximum peak value of 2.2 mV in comparison to air with a second peak value of 6.7 mV.

The refractive index of the sample is calculated using Naftlay (2014):

$$n(\omega) = 1 + \frac{c[\varphi_{\text{samp}}(\omega) - \varphi_{\text{ref}}(\omega)]}{\omega \times d_{\text{samp}}}, \tag{3.1}$$

where $\varphi_{\text{samp}}(\omega) - \varphi_{\text{ref}}(\omega)$ is the phase difference between the sample and the reference, corresponding to the shift in the time-domain, and c being the speed of light in free-space. The thickness of the sample is given by d_{samp}, which is fixed to 1 mm with the help of a spacer in this study. The absorption coefficient is calculated using the following equation Naftlay (2014):

$$\alpha(\omega) = -\frac{2}{d_{\text{samp}}} + \ln\left[\frac{E_{\text{samp}}(\omega)}{T(\omega) \times E_{\text{ref}}(\omega)}\right], \tag{3.2}$$

where $T(\omega) = \frac{4n(\omega)}{(n(\omega)+1)^2}$ is the transmission coefficient, $|E_{\text{samp}}(\omega)|$ and $|E_{\text{ref}}(\omega)|$ respectively represent the magnitudes of the sample and the reference in the frequency domain. The transfer function $H(\omega)$ is the ratio of the two defined as:

$$H(\omega) = T(\omega) \times \exp\left[-\alpha(\omega)\frac{\omega \times d_{\text{samp}}}{c}\right] \times \exp\left[-j(n(\omega) - 1)\frac{\omega \times d_{\text{samp}}}{c}\right]. \tag{3.3}$$

The results for the refractive index and absorption coefficient are illustrated in Figures 3.6 and 3.7 respectively, which show that the refractive index value decreases and the absorption coefficient increases with the frequency, which can be explained using the Kramers-Kronig relation given for dispersive materials. It is a special Hilbert transform pair and yields a comprehensive dispersion curve spanning the entire frequency spectrum Cundin and Roach (2010). Table 3.1 shows the extracted parameters at selected frequencies. The measured refractive index results are compared with Yuce et al. (2007) and Berry et al. (2003a) and are in good agreement; however, it is still a challenge to predict the nature and biological details of the samples. The samples used were either

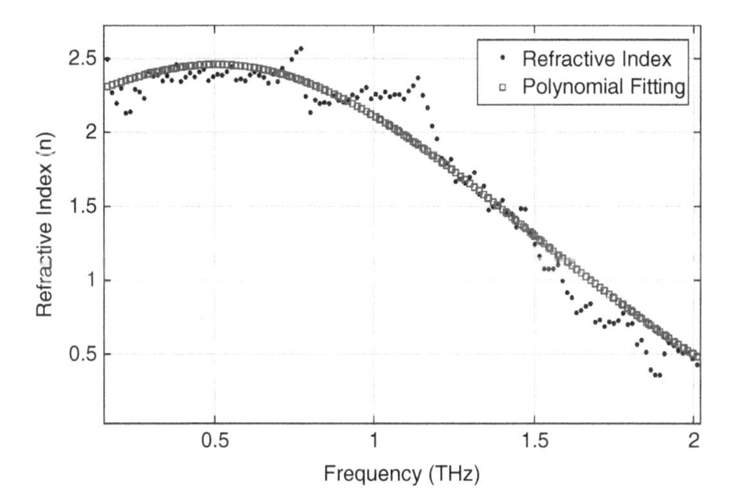

Figure 3.6 Measured refractive index as a function of frequency. The refractive index decreases with the increasing frequency from 0.8 THz to 1.2 THz.

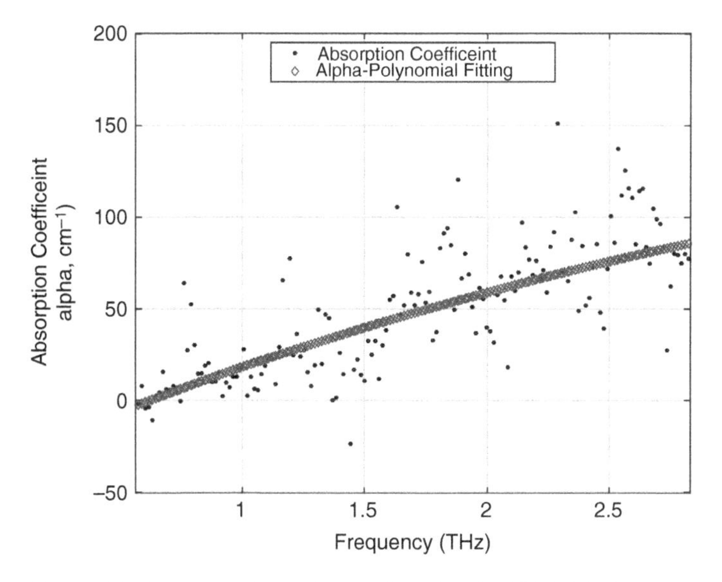

Figure 3.7 Measured absorption coefficient using THz-TDS. The alpha value is significantly low due to the dehydration and the main absorption of THz radiation as a result of atmospheric water vapours.

sliced porcine, in-vivo human skin, or frozen samples. For in-vivo skin, the measurements were done directly on the volunteers. It is important to note that skin structure is layered and intricate; therefore it is essential to include material properties of different layers of the skin. In this chapter, we deliberately measured the dermis layer of the skin and found out that at 1 THz the refractive index value is 2.1, which decreases to 1.8 at 1.2 THz. The absorption coefficient (calculated using Eq. (3.2)) values are considerably low (Fig. 3.7) due to uniform dehydration of the sample. This amounts to the fact that water in these tissues

Table 3.1 Measured material parameters using the THz-TDS technique.

Frequency (THz)	Refractive index (n)	Absorption coefficient (α)
0.8	2.33	8.96
1	2.10	18.450
1.2	1.81	27.55

plays a significant role when interacted with THz radiation. Repeatability of experimental techniques and hence the stability of obtained parameters value is investigated by repeating each measurement four times on the same sample with the thickness intact resulting in consistent refractive index value.

3.4 Tissue Sample Preparation

Rat-tail type 1 collagen, which is most commonly found in tendons, skin, ligaments, and many interstitial connective tissues, is applied as a base reagent for gel preparation. The fibroblast cells are added to fetal bovine serum (FBS), which could supply essential nutrients for cell growth. Also, the concentrated modified eagles medium (MEM), which also contains a balance of nutrients to feed the fibroblasts is added to the mixture of the collagen and fibroblast. A small quantity of sodium hydroxide is later added carefully to the mixed solution to set the gel until the pH indicator in the MEM turns pink. The gel is then incubated (5% CO_2) for gelation at 37° C for approximately 45 min. Once the initial gelation takes place, the diameter of the mixture is measured to observe any signs of contraction. After keeping the sample in an incubator for almost a week, the diameter of the gel should be measured constantly to make it contract to its maximum value. Subsequently, the gel, shown in Figure 3.8 is kept in the refrigerator to keep fresh.

3.5 Measurement Apparatus

The human skin is mainly composed of water (Mitchell et al., 1945), which leads to the high absorption of THz radiation. For better-controled experiments, dry skin is therefore preferred for measurements of glycerol-treated samples; specifically, the dermal layer remains unperturbed, when examined in standard light microscopy. When comparing to a freshly excised skin, the only change one would expect is the hydration level. Since the dry samples have been preserved for over 10 years, their hydration level is expected to be low. Hence one would only observe shrinkage of some keratinocytes (Richters et al., 1996).

The skin samples are wedged between two polymethylpentene (commonly known as TPX) slabs with a spacer of known thickness. TPX is used as a sample holder since it is transparent to THz radiation; the light is mostly transmitted through the sample with minimal

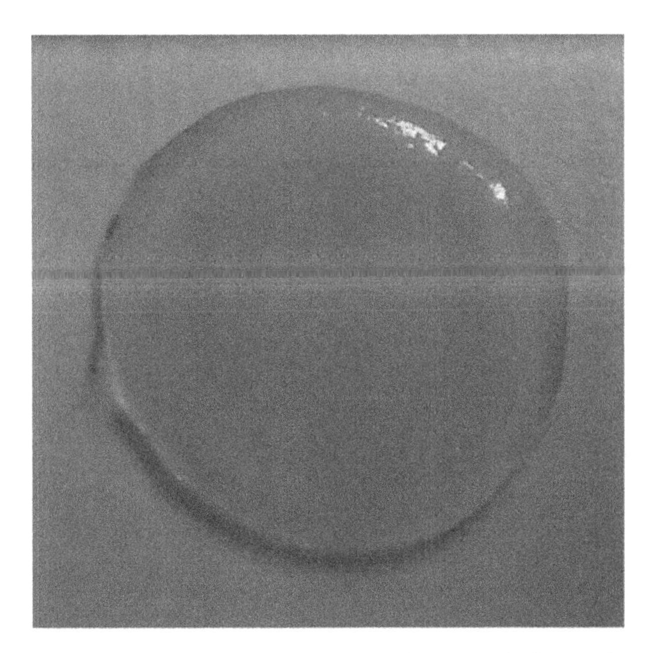

Figure 3.8 Collagen sample prepared for the study. Source: Chopra et al. (2016).

absorption in TPX (Birch and Nicol, 1984). These measurements are performed applying the aforementioned THz TDS system and therefore phase and amplitude information is collected and post-processed using transfer equation-based algorithm (Dorney et al., 2001). The TDS pulses are generated and detected via mode-locked laser. The lock-in amplifier locks and records the detected THz data for all three samples: air, TPX and skin and is aided by a computer-generated program written in LabView®.

A schematic illustration of the setup of a THz-TDS setup is shown in Figure 3.9, where the transmission mode is used. A titanium-sapphire pulsed laser produces femtosecond pulses, which are then redirected into two separate optical paths by a beam-splitter. One beam becomes the receiver pulse (a measurement using the receiver is possible only when a femtosecond pulse is incident on the receiver), while the other is used to excite a THz wave at the emitter. Parabolic mirrors are used to focus and collimate the THz pulses travelling through the sample and onto the detector. The detector is activated only when it is struck by a femtosecond pulse from the detection beam-line. When the detector is photo-excited, the electric field of the THz pulse causes an electric current to flow inside the detector. The time delay on the detector beam-line is used to control the relative time delay between the THz pulses. When the detector is conducting, the electric field of the THz pulses can be plotted over time.

After the collagen sample is cultivated, it is stored in an icebox for measurements. During the experiment, the system without the holder is measured first to get the time-domain pulse response of the air and then the response of the system with empty holder, made of TPX with low absorption ($\leq 1\text{cm}^{-1}$) and almost constant refractive index (1.46) over the frequency range of interest. Finally, the holder with the sample is measured. For each step, the measurement is repeated 3 times to obtain a stable value; also, for each sample with

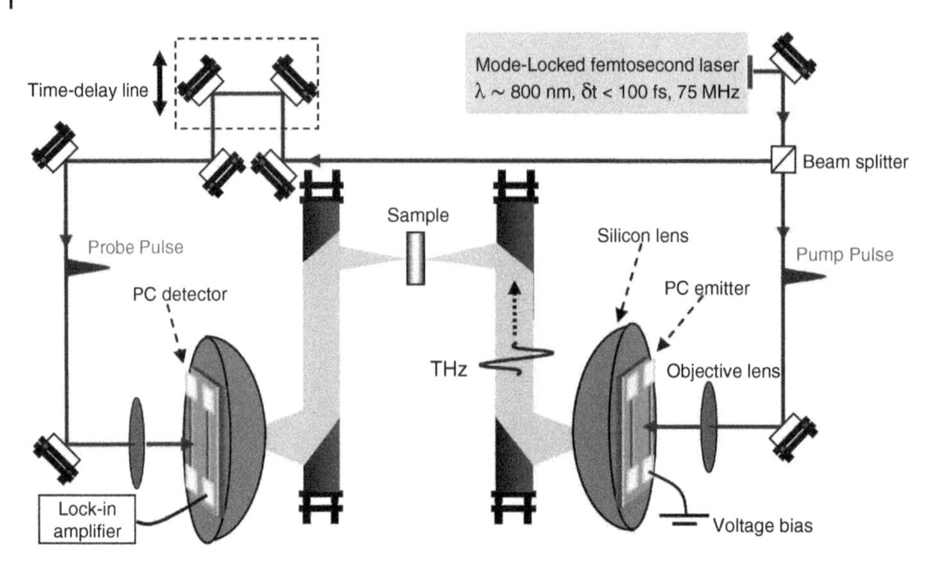

Figure 3.9 Schematic diagram of a THz-TDS system operating in the transmission mode.

different thickness, three measurements are conducted and the mean value of the final results is adopted. Dry air or nitrogen is pumped into the system to eliminate the effects of the vapour.

3.6 Simulating the Human Skin

The human skin is a collection of complex, heterogeneous and anisotropic materials, in which secondary components like blood vessel and pigment content are spatially distributed in depth (Montagna, 2012). The skin tissues can be represented by a three-layer model (i.e., SC, epidermis, and dermis. The dimensions are shown in Fig. 3.10). In the skin model, the roughness of boundaries between the SC and epidermis can be considered as the order of magnitude for THz wavelength (Yang et al., 2016). Since sweat glands are distributed almost all over the human body and represent a form of cooling in humans, therefore, their consideration in the model was essential to get a closer insight of EM wave propagation inside the real human tissue and their effect on the signal propagation losses through the skin. The optical coherence tomography imaging (Hayut et al., 2013) of human skin shows that the sweat ducts are of spiral shape. The detailed model of human skin with sweat duct was initially presented by Hayut et. al. in (Hayut et al., 2013), where the sweat duct was modelled in the epidermis layer (Hayut et al., 2013) by a helix (Yang et al., 2016) as shown in Figure 3.10 with a height of $265\mu m$ and diameter equal to $40\mu m$. As the duct becomes only electrically conductive only when it contains sweat (with a composition of 99% water and 1% salt and amino acid) (Shafirstein and Moros, 2011), therefore, the permittivity of water in the THz band can be used to describe the material properties of the sweat duct. In real skin tissue, the thickness of each layer is highly dependant on the location on the body, ranging from very thin such as eyelids, to very thick when covering the tough areas, such as the hand palm. The skin tissue dimensions

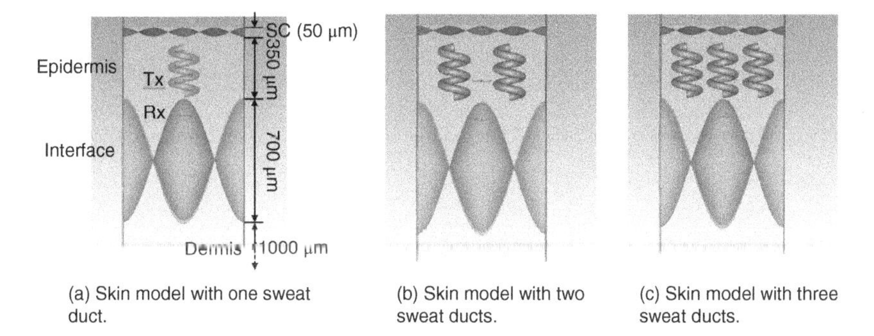

(a) Skin model with one sweat duct.

(b) Skin model with two sweat ducts.

(c) Skin model with three sweat ducts.

Figure 3.10 Numerical skin model (based on CST Microwave Studio™) representing the layers and their thickness, while including the different numbers of sweat ducts (dimensions for each layer are shown as well. (Source: Yang et al., 2016)).

in the model were taken to be the average between the skin dimensions of the sensitive areas and the tough areas. Two dipoles optimized to ensure that the impedance matching is better than -10 dB (modelled as nano-antennas working in the THz frequency band) were used as transmit (Tx) and receiver (Rx) antennas. In the simulation, one antenna was placed in the epidermis while the other was located in the dermis layer. The transmission coefficient was recorded in the frequency range of 0.8 - 1.2 THz. Figure 3.10 represents the proposed skin layer model having a different spacing between the Tx and Rx antennas by varying number of sweat ducts. All the simulations were carried out using CST Microwave Studio™ on a high-performance computing cluster.

3.6.1 Human Body Channel Modelling

The Friis transmission equation is commonly used to estimate the power received by an antenna. Recently, the formula has been modified by Jornet et. al. (Jornet and Akyildiz, 2011), in which the path loss of a THz channel inside a human skin is derived. The path loss can be divided into two parts, the spreading path loss PL_{spr}, which is due to the expansion of waves inside the tissue and the absorption path loss PL_{abs} (due to absorption of waves in the tissue). Similarly, the path loss in human tissues can also be divided into two parts,

$$\begin{aligned} \mathrm{PL}_{\mathrm{total}}[dB] &= \mathrm{PL}_{\mathrm{spr}}(f,d)[dB] + \mathrm{PL}_{\mathrm{abs}}(f,d)[dB] \\ &= 20\log\frac{4\pi d}{\lambda_g} + 4.342\alpha \times d, \end{aligned} \tag{3.4}$$

where f represents the frequency, d is the path length, α is the extinction coefficient that measures the amount of absorption loss and defined as $\frac{4\pi K}{\lambda}$ where K is the propagation constant. This path loss model takes into account only the distance and frequency and it does not explicitly discuss losses inside the skin.

It is important to investigate the correlation between any measurement data and explanatory variables in the environment. Only then, a reliable path loss model can be deduced from the measurements. Factors such as distance, the number of ducts and frequency have to be incorporated into the model. The relationship between distance and measured path loss varies with distance model and different models are investigated to find out the model

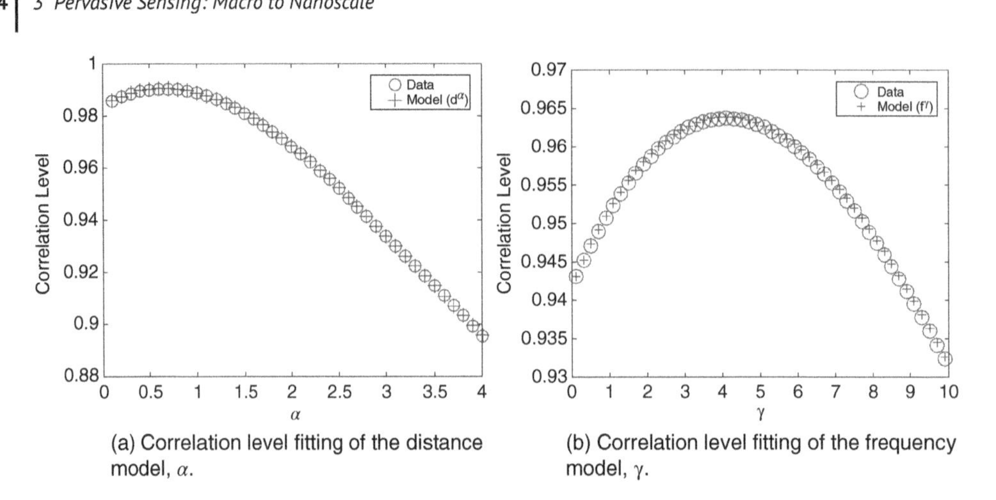

(a) Correlation level fitting of the distance model, α.

(b) Correlation level fitting of the frequency model, γ.

Figure 3.11 Correlation level fitting of eq. (3.5) and eq. (3.6), used to calculate the value of α (distance model) and γ (frequency model), which is the maximum value in fitting.

that has the highest correlation level. This indicates that the highest level of information captured by the model from measured path loss. As a first step, the correlation with distance in a logarithmic scale is tested. The correlation level between path loss (PL) in dB scale and distance in the logarithmic domain is 0.98. To find out whether the PL follows other distance models, the correlation of measured PL with d^{α} is investigated. Figure 3.11a shows the variation of correlation level with the exponent of distance model α. The results in Figure 3.11a show that there are some values of α, where the distance model d_{α} has a higher correlation level than the logarithmic distance model, i.e. $\log_{10}(d)$. To find the value of α that has the highest correlation level, a model of correlation level (C) has been inferred from correlation level data as a function of α,

$$C(\alpha) = 0.00181\alpha^3 - 0.018\alpha^2 + 0.021\alpha + 0.984. \tag{3.5}$$

The distance model that has the highest correlation is the model that has α, which fulfils $\frac{dC}{d\alpha} = 0$. The solution of this equation is $\alpha = 0.65$, which leads to a correlation value of 0.9903. The correlation level fitting of Eq. (3.5) is shown in Figure 3.11a. An interim distance model ($d^{0.65}$) is used to extract the information of PL data that are not a function of the distance model as the error between the PL and the interim distance model. Then, this extracted residual is tested against the frequency as an explanatory model variable to check, if it has any correlation. This correlation between the residual of PL after excluding the effect of the distance model and frequency model is tested against frequency models of $\log_{10}(f)$ and f^{γ}. The correlation level with $\log_{10}(f)$ is 0.94 and with frequency models, different values of γ are shown in Figure 3.11a, from which it is clear that there are some values of γ, where the frequency model has a higher correlation than the model $\log_{10}(f)$. To find the value that has the highest correlation value, the correlation level is modelled as a function of γ,

$$C_f(\gamma) = 3.97 \times 10^{-5}\gamma^3 - 0.00164\gamma^2 + 0.0114\gamma + 0.942. \tag{3.6}$$

The value of γ, which gives highest correlation for frequency model is obtained from $\frac{dC_f}{d\gamma} = 0$, which is $\gamma = 4.07$. To find a multivariate PL model, a multivariable regression technique has been invoked. This is to find the relationship between two or more explanatory variables by fitting a linear equation to the collected data. After intensive experiments, it turns out that the model should be of the form,

$$PL = A(N) + B(N)d^{0.65} + C(N)f^{4.07}, \tag{3.7}$$

where the regression technique is used to find the functions $A(N)$, $B(N)$, and $C(N)$, which are constant offset coefficients of distance and frequency as functions of the number of ducts, respectively. The modelling process is conducted in two steps: (i) the fitting process is done for fixed values of N, and model fitting is performed for corresponding data of PL; (ii) a series of data for A, B and C corresponding to various values of N, is modelled as affine functions of N. The fitting process is based on least-squares fit for a particular value of N, which can be formulated as,

$$\mathbf{PL} = \boldsymbol{X\beta}, \tag{3.8}$$

$$\boldsymbol{X} = \begin{bmatrix} 1 & d_1^{0.65} & f_1^{4.07} \\ 1 & d_2^{0.65} & f_2^{4.07} \\ \vdots & \vdots & \vdots \\ 1 & d_3^{0.65} & f_3^{4.07} \end{bmatrix} \tag{3.9}$$

$$\boldsymbol{\beta} = \begin{bmatrix} A1 \\ B1 \\ C1 \end{bmatrix} \tag{3.10}$$

$$\mathbf{PL} = \begin{bmatrix} \mathrm{PL}_1 \\ \mathrm{PL}_2 \\ \vdots \\ \mathrm{PL}_n \end{bmatrix} \tag{3.11}$$

The goal here is to find the best values of A_i, B_i and C_i for each value of $N = i$, that minimizes the difference between left and right sides such that:

$$\hat{\beta} = min_{\beta} S(\beta), \tag{3.12}$$

$$S(\beta) \parallel \mathrm{PL} - X\beta \parallel . \tag{3.13}$$

S is the objective function that has to be solved using a quadratic minimization problem since Eq. (3.7) has no solution. The solution of this optimization problem (Eq. (3.13)) is well-known and is given as (Kay, 1993):

$$\hat{\beta} = (\mathbf{X}^T\mathbf{X})^{-1}\mathbf{X}^T\ \mathbf{PL}. \tag{3.14}$$

Each of the coefficients (A_i, B_i and C_i) for different values of N is fitted with affine function of N,

$$A(N) = a_1 + b_1 N, \tag{3.15}$$

$$B(N) = a_2 + b_2 N, \tag{3.16}$$

$$C(N) = a_3 + b_3 N. \tag{3.17}$$

Table 3.2 Fitting statistics Of the proposed model with respect to the measurement.

Correlation Level	R-squared	Adjusted R-squared	RMSE
0.9993	0.9987	0.9987	0.3407

The final form of the path loss model inferred from measurement data after the two described fitting process is,

$$\mathbf{PL} = -0.2 \times N + 3.98 + (0.44 \times N + 98.48)d^{0.65} + (0.068 \times N + 2.4)f^{4.07}, \tag{3.18}$$

where, f is the frequency in THz, d is the distance in mm, and N is the number of sweat ducts, respectively.

The fitting statistics of the proposed model are shown in Table 3.2, where methods such as the r-squared, and the coefficient of determination were used in which the closeness of the statistical data to the measured data is gauged. However, the r-squared method does not determine whether the estimates of the prediction coefficients are biased or not. It may also mislead the goodness of the fit as a result of overfitting process. These problems are addressed in the adjusted r-squared measure which adjusts the number of explanatory variables to the measurement data. This is an important measure in the multi-variable linear regression since it penalizes the model for adding more nonsense variables. The root-mean-squared error (RMSE) is an absolute measure of how close the model is to the measurement data. It can be interpreted as the standard deviation of unexplained and un-modelled variation in data. A lower value of RMSE corresponds to a better fit.

3.7 Networking and Communication Mechanisms for Body-Centric Wireless Nano-Networks

In Figure 3.12, the illustrated nano-nodes are the simplest and smallest nano-devices to perform simple computation and detection, and then transmit it to relay nodes, which are also small nano-devices. Their task is to amplify and forward the received signal to nano-routers. These nano-devices (nano-routers) are slightly larger in terms of the computational and behaviour capabilities and can also act as a control unit for the set of nano-nodes by ordering simple commands like reading, sleep, wakeup etc. They can be invasive or non-invasive depending on the application. The nano-micro interface is composed of hybrid devices which are used to exchange information between the two interfaces and the last unit i.e., the gateway allows user to remotely control the system using the internet.

Since the path loss at terahertz frequencies is considerably large even at a very small distance and to transmit information to relatively larger distance, multiple relays have to be used to assist communication between the transmitter nano-nodes and the nano-router/Rx node as shown in Figure 3.13. To ensure a simplistic implementation of the relays, an amplify-and-forward relaying assumption has been made.

Due to the high pathloss, a direct Tx-Rx link would result in negligible received signal-to-noise ratio (SNR) at the Rx and thus it can be ignored. The received SNR can be

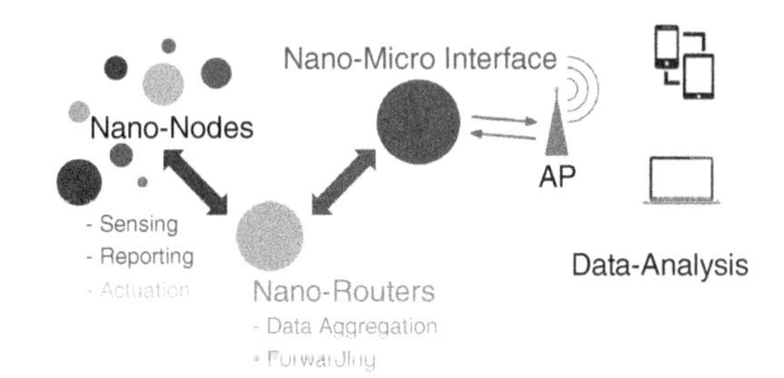

Figure 3.12 Envisioned architecture for nano-healthcare.

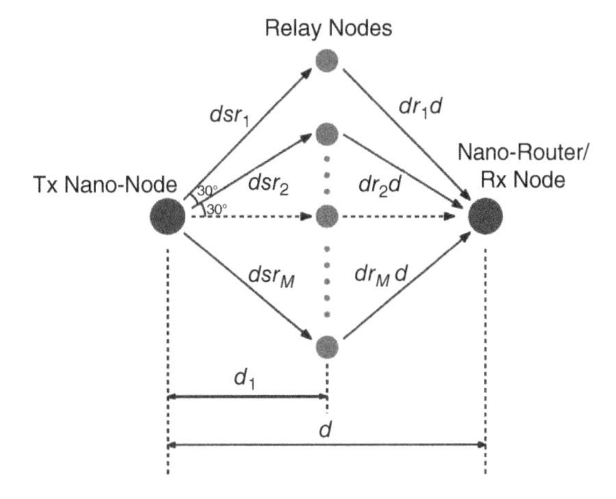

Figure 3.13 System model for in-vivo cooperative communication at terahertz frequencies.

optimized by employing a maximum ratio combining at the Rx side i.e. as a nano-router (Goldsmith, 2005). As a result, the received SNR at the Rx, γ_{RX}, through relaying link is given by (Soliman and Beaulieu, 2012),

$$\gamma_{RX} = \sum_{i=1}^{M} \frac{\gamma_{s,r_i}\gamma_{r_i,d}}{\gamma_{s,r_i} + \gamma_{r_i,d} + 1}, \tag{3.19}$$

where $\gamma_{s,r_i} \triangleq \frac{|h_{s,r_i}|^2 P_s}{\ell_{s,r_i}\sigma^2}$ is the received SNR at the ith relay due to the link between transmitter and the ith relay, $\gamma_{r_i,d} \triangleq \frac{|h_{r_i,d}|^2 P_r}{\ell_{r_i,d}\sigma^2}$ is the received SNR at the destination node due to the link between the ith relay and the Rx, ℓ_{s,r_i} and $\ell_{r_i,d}$ are the path losses from source transmitter to the ith relay and from the ith relay to the destination or Rx, respectively, $h_{s,r_i} \sim \mathcal{CN}(0,1)$ and $h_{r_i,d} \sim \mathcal{CN}(0,1)$ are complex normally distributed channel co-efficients for the respective links, σ^2 is the variance of additive white Gaussian noise, P_s is the transmit power of source (transmitter) and P_{r_i} is the transmit power of the ith relay. We proposed a pathloss model, ℓ_{s,r_i} or $\ell_{r_i,d}$ at the THz frequency (Abbasi et al., 2016), which is can be written as,

$$\ell_{s,r_i} = -0.2N + 3.98 + (0.44N + 98.48)d_{s,r_i}^{0.65} + (0.068N + 2.4)f^{4.07} \tag{3.20a}$$

$$\ell_{r_i,d} = -0.2N + 3.98 + (0.44N + 98.48)d_{r_i,d}^{0.65} + (0.068N + 2.4)f^{4.07}, \tag{3.20b}$$

where $N = 5$ is the number of sweat ducts, f is frequency and d_{s,r_i} and $d_{r_i,d}$ are the Euclidean distances from source transmitter to the ith relay and from the ith relay to the destination or Rx, respectively.

3.8 Concluding Remarks

In this chapter, an overview of the field of on-body communication at terahertz frequencies has been presented along with some of the basic theory of the human skin characteristics, and some of the underpinning concepts governing the electromagnetic wave propagation. Some of the important challenges faced in on-body communication were discussed through the presentation of path loss models. The detailed material composition of the human skin was also shown. Furthermore, an overview was provided to perform material characterization measurements of the human skin using the terahertz time-domain spectroscopy technique. This work serves as an important launchpad to stimulate further research on reliable and energy-efficient communication protocols for THz in-vivo systems.

References

Abbasi, Q.H., H. El Sallabi, N. Chopra, K. Yang, K.A. Qaraqe, and A. Alomainy. (2016) “Terahertz channel characterization inside the human skin for nano-scale body-centric networks,” *IEEE Transactions on Terahertz Science and Technology*, 6(3), p. 427–434.

Akyildiz, I.F., and Josep Miquel Jornet. (2010a) “Electromagnetic wireless nanosensor networks,” *Nano Communication Networks*, 1(1), p. 3–19.

Akyildiz, I.F., and J.M. Jornet. (2010b) “Electromagnetic wireless nanosensor networks,” *Nano Communication Networks*, 1(1), p. 3–19.

Akyildiz, I.F., F. Brunetti, and C. Blázquez. (2008a) “Nanonetworks: a new communication paradigm,” *Computer Networks*, 52(12), p. 2260–2279.

Akyildiz, I.F., F. Brunetti, and C. Blázquez. (2008b) “Nanonetworks: a new communication paradigm,” *Computer Networks*, 52(12), p. 2260–2279.

Berry, E., A.J. Fitzgerald, N.N. Zinov’ev, G.C. Walker, S. Homer-Vanniasinkam, C.D. Sudworth, R.E. Miles, J.M. Chamberlain, and M.A. Smith. (2003a) “Optical properties of tissue measured using terahertz pulsed imaging,” *Proceedings of SPIE: Medical Imaging 2003: Physics of Medical Imaging*, 5030, p. 459–470.

Berry, E., A.J. Fitzgerald, N.N. Zinov’ev, G.C. Walker, S. Homer-Vanniasinkam, C.D. Sudworth, R.E. Miles, J.M. Chamberlain, and M.A. Smith. Optical properties of tissue measured using terahertz-pulsed imaging. (2003b) in Yaffe, M.J., and Antonuk, L.E. (eds,), *Medical imaging 2003: physics of medical imaging*, vol. 5030. Bellingham, WA: International Society for Optics and Photonics, p. 459–470.

Binzoni, T., A. Vogel, A.H. Gandjbakhche, and R. Marchesini. (2008) “Detection limits of multi-spectral optical imaging under the skin surface,” *Physics in medicine and biology*, 53(3), p. 617.

Birch, J.R., and E.A. Nicol. (1984) “The FIR optical constants of the polymer TPX,” *Infrared physics*, 24(6), p. 573–575.

Blanpain, C., and E. Fuchs. (2009) “Epidermal homeostasis: a balancing act of stem cells in the skin,” *Nature Reviews Molecular Cell Biology*, 10(3), p. 207–217.

Chan, W.L., J. Deibel, and D.M. Mittleman. (2009) “Imaging with terahertz radiation,” *Reports on Progress in Physics*, 70(8), p. 1325.

Chopra, N., K. Yang, J. Upton, Q.H. Abbasi, K. Qaraqe, M. Philpott, and A. Alomainy. (2016) "Fibroblasts cell number density based human skin characterization at THZ for in-body nanonetworks," *Nano Communication Networks*, 10, p. 60–67.

Cundin, L.X., and W.P. Roach. (2010) "Kramers-kronig analysis of biological skin," *arXiv preprint arXiv:1010.3752.*

Dorney, T.D., R.G. Baraniuk, and D.M. Mittleman. (2001) "Material parameter estimation with terahertz time-domain spectroscopy," *JOSA A*, 18 (7), p. 1562–1571.

Feldman, Y., A. Puzenko, P.B. Ishai, A. Caduff, I. Davidovich, F. Sakran, and A.J. Agranat. (2009) "The electromagnetic response of human skin in the millimetre and submillimetre wave range," *Physics in Medicine & Biology*, 54 (11), p.3341.

Fitzgerald, A.J., E. Berry, N.N. Zinov'ev, S. Homer-Vanniasinkam, R.E. Miles, J.M. Chamberlain, and M.A. Smith. (2003) "Catalogue of human tissue optical properties at terahertz frequencies," *Journal of Biological Physics*, 29 (2), p. 123–128, ISSN 1573-0689.

Gabriel, C., S. Gabriel, and E. Corthout. (1996) "The dielectric properties of biological tissues: I. literature survey," *Physics in Medicine and Biology*, 41 (11), p. 2231–2249.

Gallerano, G.P. (2004) "Tera-Hertz radiation in biological research, investigations on diagnostics and study on potential genotoxic effects," Technical report, THz-BRIDGE, 05 2004.

Gawkrodger, D., and M.R. Ardern-Jones. *Dermatology e-book: an illustrated colour text.* https://www.elsevier.com/books/dermatology/gawkrodger/978-0-7020-6849-2. Elsevier Health Sciences.

Goldsmith, A. (2005) *Wireless communications.* New York: Cambridge University Press.

Grice, E.A., and J.A. Segre. (2001) "The skin microbiome," *Nature Reviews Microbiology*, 9(4), p. 244–253.

Hayut, I., A. Puzenko, P.B. Ishai, A. Polsman, A.J. Agranat, and Y. Feldman. (2013) "The helical structure of sweat ducts: their influence on the electromagnetic reflection spectrum of the skin," *IEEE Transactions on Terahertz Science and Technology*, 3(2), 207–215.

Hey, A. (2018) *Feynman and computation.* Boca Raton, FL: CRC Press, 2018.

Hintzsche, H., C. Jastrow, T. Kleine-Ostmann, U. Kärst, T. Schrader, and H. Stopper. (2012) "Terahertz electromagnetic fields (0.106 thz) do not induce manifest genomic damage in vitro," *PLOS ONE*, 7(9), p. 1–8.

IEEE. (2016) "IEEE recommended practice for nanoscale and molecular communication framework," *IEEE Std 1906.1-2015*, p. 1–64, January.

Jornet, J.M., and I.F. Akyildiz. (2010) "Graphene-based nano-antennas for electromagnetic nanocommunications in the terahertz band." In *Proceedings of the Fourth European Conference on Antennas and Propagation*, April p. 1–5.

Jornet, J.M., and I.F. Akyildiz. (2011) "Channel modeling and capacity analysis for electromagnetic wireless nanonetworks in the terahertz band," *IEEE Transactions on Wireless Communications*, 10(10), p. 3211–3221, October.

Jornet, J.M., J.C. Pujol, and J.S. Pareta. (2012) "PHLAME: A physical layer aware MAC protocol for electromagnetic nanonetworks in the terahertz band," *Nano Communication Networks*, 3(1), p. 74–81.

Kay, S.M., *Fundamentals of statistical signal processing.* Upper Saddle River, NJ: Prentice Hall PTR, 1993.

Kelly, M.J. (2013) *Manufacturability and nanoelectronic performance*. Hoboken, NJ: John Wiley & Sons, Inc., 133–138.

Latré, B., G. Vermeeren, L. Martens, and P. Demeester. (2004) "Networking and propagation issues in body area networks." Proceedings of the 11th Symposium on Communications and Vehicular Technology in the Benelux, SCVT 2004.

Mitchell, H.H., T.S. Hamilton, F.R. Steggerda, and H.W. Bean. (1945) The chemical composition of the adult human body and its bearing on the biochemistry of growth, *Journal of Biological Chemistry*, 158(3), p. 625–637.

Montagna, W. (2012) *The structure and function of skin*. https://www.elsevier.com/books/the-structure-and-function-of-skin/montagna/978-0-12-505263-4. Elsevier, 2012.

Naftlay, M. (2014) *Teraherz metrology*. Boston and London: Artech House.

Pickwell, E., and V.P. Wallace (2006) Biomedical applications of terahertz technology. *Journal of Physics D: Applied Physics*, 39(17), p.R301.

Pickwell, E., B.E. Cole, A.J. Fitzgerald, M. Pepper, and V.P. Wallace (2004) In vivo study of human skin using pulsed terahertz radiation, *Physics in Medicine and Biology*, 49(9), p. 1595.

Piro, G., K. Yang, G. Boggia, N. Chopra, L.A. Grieco, and A. Alomainy (2015a) Terahertz communications in human tissues at the nanoscale for healthcare applications, *IEEE Transactions on Nanotechnology*, 14(3), p. 404.

Piro, G., K. Yang, G. Boggia, N. Chopra, L.A. Grieco, and A. Alomainy (2015b) Terahertz communications in human tissues at the nanoscale for healthcare applications, *IEEE Transactions on Nanotechnology*, 14(3), p. 404.

Richters, C.D., M.J. Hoekstra, J. Van Baare, J.S. Du Pont, and E.W.A. Kamperdijk (1996) Morphology of glycerol-preserved human cadaver skin, *Burns*, 22(2), p. 113–116.

Shafirstein, G., and E.G. Moros (2011) Modelling millimetre wave propagation and absorption in a high resolution skin model: the effect of sweat glands, *Physics in medicine and biology*, 56(5), p. 1329.

Shnayder, V., B. Chen, K. Lorincz, T.R.F. Fulford Jones, and M. Welsh (2005) Sensor networks for medical care, in Proceedings of the 3rd International Conference on Embedded Networked Sensor Systems, SenSys '05. New York: ACM, p. 314.

Smye, S.W., J.M. Chamberlain, A.J. Fitzgerald, and E. Berry (2001) The interaction between terahertz radiation and biological tissue, *Physics in Medicine & Biology*, 46(9), p. R101.

Soliman, S.S., and N.C. Beaulieu (2012) Exact analysis of dual-hop af maximum end-to-end snr relay selection, *IEEE Transactions on Communications*, 60(8), p. 2135–2145.

Yang, K., A. Pellegrini, M.O. Munoz, A. Brizzi, A. Alomainy, and Y. Hao (2015) Numerical analysis and characterization of THZ propagation channel for body-centric nano-communications, *IEEE Transactions on Terahertz Science and Technology*, 5(3), p. 419–426.

Yang, K., Q.H. Abbasi, N. Chopra, M. Munoz, Y. Hao, and A. Alomainy (2016) Effects of non-flat interfaces in human skin tissues on the in-vivo tera-hertz communication channel, *Nano Communication Networks*, 8, p. 16–24.

Yuce, M.R., S.W.P. Ng, N.L. Myo, J.Y. Khan, and W. Liu (2007) Wireless body sensor network using medical implant band, *Journal of Medical Systems*, 31 (6), p. 467–474.

4

Biointegrated Implantable Brain Devices

Rupam Das and Hadi Heidari

Electronic and Nanoscale Engineering, University of Glasgow, Glasgow, United Kingdom

Implantable neural devices are the most widely applied tools in neuroscience research as well as neuroprosthetics to record neural signal activities at single-neuron and sub-millisecond resolution. However, lack of mechanical compatibility between the traditional rigid probes and brain tissues causes neuroinflammatory responses and deteriorate recorded signal quality. Decreasing the cross-sectional area while improving flexibility of the probes can significantly enhance the chronic stability of neural interfaces. Additionally, magnetic resonance imaging (MRI) compatibility is another issue that should be addressed for the development of future implantable devices. Herein, current advancement in developing the chronically stable implantable neural devices is highlighted, with a emphasis on making use of the sophisticated materials and structural design concepts.

4.1 Background

As a complex organ, the brain controls the center of our body and directly affects every operation, starting from emotions, rational thinking, breathing, heartbeat, sleep, food and fluid intake, and so on. A healthy brain is essential for quality of life and sustainable well-being. However, for brain disorders (Figure 4.1), neurological and mental alike, these numbers are increasing steadily. According to the European Brain Council, with yearly costs of about 800 billion euros and an estimated 179 million people afflicted in 2010, brain diseases are an unquestionable emergency and a grand challenge for neuroscientists (Gustavsson et al. 2011). Despite recent successes in medical science and recent breakthroughs which have empowered us with the dual possibility to ease the societal burden of mental and brain diseases and innovate at the frontiers of technology, the intrinsic complexity of the brain has hampered our translational capacity, indicating that more insight is needed to treat neurological disorders efficiently.

Over the past few decades, state-of-the-art technologies such as Electroencephalography (EEG), Electrocorticography (ECoG), and electrical stimulation based on the implantable neural devices (neural implants or probe) have provided tremendous insight into the brain's functionality (Seymour et al. 2017). Among these technologies, electrical or focal

Engineering and Technology for Healthcare, First Edition.
Edited by Muhammad Ali Imran, Rami Ghannam and Qammer H. Abbasi.

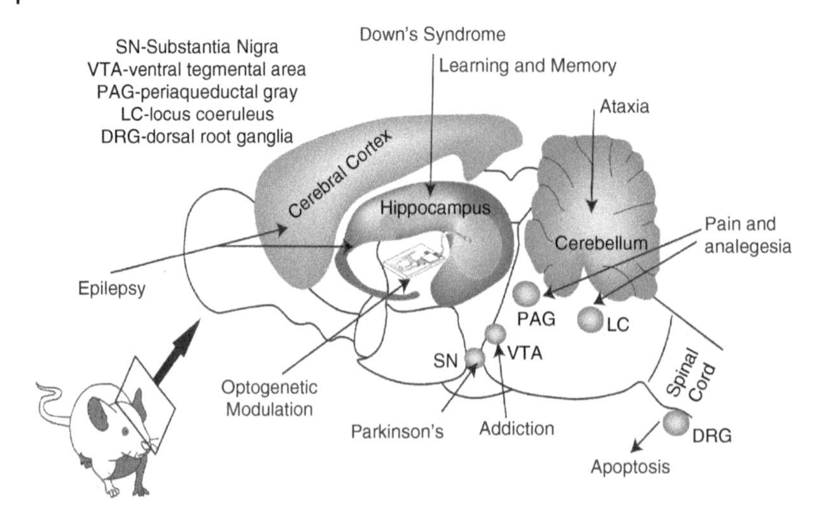

Figure 4.1 Diagram of Sagittal View of a Rat Brain Indicating Regions Which Are Believed to Play a Role in Neurological Diseases.

brain stimulation based on implantable devices remains a therapeutic strategy of interest for people with medication-resistant forms of epilepsy and who are not candidates for surgery (Terry 2009). However, shortness of breath, throat pain, and cough can be caused by the undesired electrical stimulation of neurons, thereby restricting the extent of this approach (Chou et al. 2012). Optogenetics is another neural modulation technique which utilizes light to stimulate genetically engineered neurons, providing a better option for controlling the cells compared to conventional electrical stimulation. Recently, optogenetics has provided controlled stimulation using photosensitive ion channels or proteins (e.g., Channelrhodopsin-2 (ChR2)) in genetically modified neurons to allow optical stimulation (470 nm blue light) or inhibition (580 nm yellow light) (Deisseroth 2011). Nevertheless, scientists pursuing optogenetics treatments for brain diseases still face some technical challenges, for example, traditional optogenetics methods for powering the neural implants relies on stiff and tethered (e.g., optical fibres) systems (Gutruf and Rogers 2018). Furthermore, implant-tissue biointegration is vital to cause minimum tissue damage and negligible immune response for chronic reliability (Gutruf and Rogers 2018; Checa et al. 2019; Zuo et al. 2020 ; Nabaei et al. 2020). At the same time, an encounter between an implant and the magnetic resonance imaging (MRI) environment introduces complications such as exertion of force, heating, voltage induction and imaging artefacts (Das and Yoo 2017). As a result, MRI-compatible biointegrated chronic in vivo optogenetics systems permit anatomical and functional MRI (fMRI) researches across the whole brain without any intervention with the MRI systems (Zhao et al. 2016).

This chapter discusses unique design methods to develop flexible, soft, and biointegrated interfaces for brain implantable devices by revealing mechanical and physical stability with nerual tissues. The discussion begins by introducing different neural interfaces, considering crucial parameters for implantable neural devices. The following section features the progress and challenges in biointegration of implantable neural probes. Section 4.4 highlights the MRI compatibility of neural implants, and we also discuss cutting-edge advancement in implantable neural interfaces and suggest the scope for future studies. We finally conclude the chapter in section 4.5.

4.2 Neural Device Interfaces

Depending on the location of the collected signal, neural interfaces may be classified as electroencephalography (EEG), electrocorticogram (ECoG), LFPs, and action potential (Seymour et al. 2017), as shown in Figure 4.2. Among these, the most basic, noninvasive method to collect neruonal activity is EEG (scalp recordings), which nowadays has been used in treating of seizure or epilepsy (Smith 2005). EEG makes it possible to observe sleep, gives better insight of psychological function of the brain and language perception (Kutas and Federmeier 2000). However, EEG cannot record nearby data from a specific brain region, and offers low transfer rates such as 5-25 bit/s (Birbaumer 2006). Additionally, recorded EEG data weakened due to lossy and dense brain tissue, as well as other brain layers such as skin and cranium, thus restricting spatiotemporal resolution. On the other hand, recent studies have focused on applying the invasive brain machine interface such as ECoG (epidural/subdural recordings). In comparison with the EEG, ECoG minimizes unwanted interferences and makes it possible to collect high frequency and high accuracy brain signals. ECoG electrodes are inserted within the cortex, thereby reducing tissue interference between the neurons and the electrodes (Rivnay et al. 2017). Nevertheless, the neural signals using ECoG are from superficial locales of the brain and do not make it possible to record data from single neurons. Recording accurate signals from single neurons with high spatiotemporal resolution over a particular region of a brain is vital to promote a more far-reaching grasp of the cognitive and sensory system. Inevitably, implantable neural devices will be necessary, which are a more invasive technique to record the LFP's signal from the deep brain region. The recording of LFPs signifies local neural activities which are obtained from specific neuronal densities and comprise action potentials and additional membrane potential fluctuations, and provide noteworthy details about the measured brain area (Lee et al. 2019). Ongoing studies indicating single-neuron recording allows us to model and decipher the wiring of the brain as well

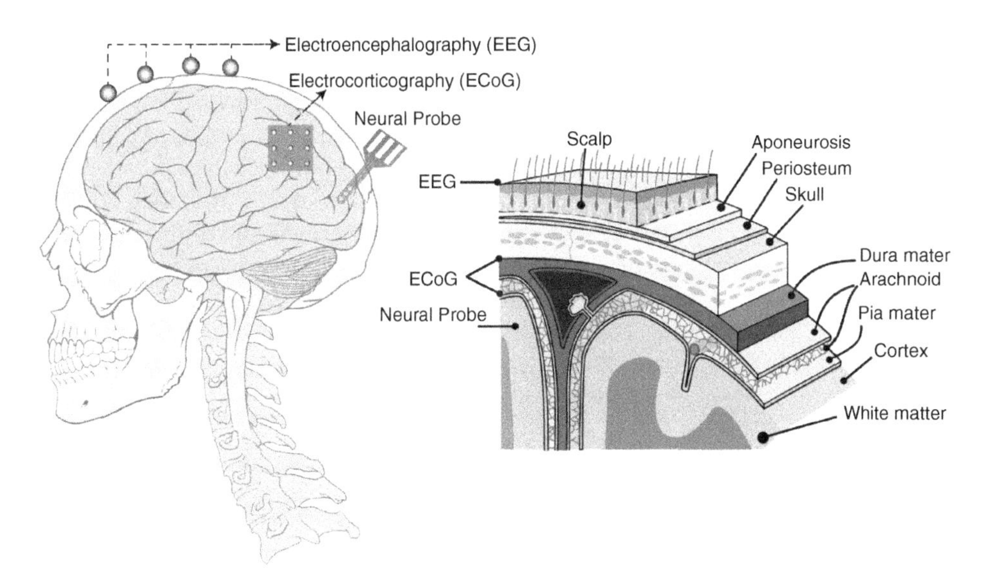

Figure 4.2 Examples of Different Neural Recording Systems.

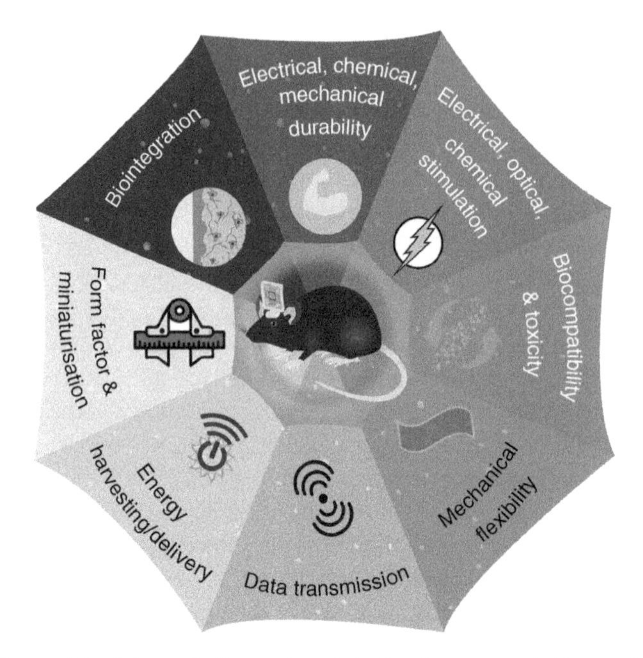

Figure 4.3 Schematic Illustration of the Key Parameters in Neuroscience Research for Implantable Devices.

as its link to movement, discernment, and memory. Recently, the implantable brain probe has found application in confining epileptogenic regions (Proctor et al. 2018) and treating Parkinson's disease (Swann et al. 2018). Among several neural interfaces, the implantable neural probe is regarded as producing the most useful control signals. These findings call for more innovations in implantable neural device technologies to facilitate high resolution, multiplexed functionality, and spatiotemporal span for neural engineering.

Current success in the area of stimulation types, materials science, mechanical design, and system engineering can make possible chronic in vivo neural recordings by applying implantable neural devices (Hong and Lieber 2018; Gutruf and Rogers 2018). Figure 4.3 indicates major design variables for developing implantable brain devices. Application of biocompatible material ensures a stable neural interface both mechanically and chemically, as well as allowing a long-lasting, least invasive operation of the brain (Hwang et al. 2012). Apart from biocompatibility, a flexible and mechanically compliant neural interface to the desired tissues is critical for chronic biointegration (Jeong et al. 2015). Additionally, to establish a tether-free neural interface for neural recordings in naturalistic cases, wireless power transfer (WPT)plays an important role (Balasubramaniam et al. 2018; Yuan et al. 2020; Das and Heidari 2019; Zhao et al. 2019a,b). As a result, a critical goal is to formulate flexible, fully wireless, miniaturized, biointegrated, and implantable platforms.

4.3 Implant Tissue Biointegration

At present, highly developed devices like Utah arrays and Michigan-style probes are available commercially and are key tools in research on neuroscience (Seymour et al. 2017).

Additionally, innovative silicon-based neural devices like neuropixels for high-density neural recordings (Jun and et al 2017), multifunctional probe (Shin and et al. 2019), and 3D probe for recording of coordinated brain activity from large population of neurons Rios et al. (2016) have enlightened us with better understanding of the brain. Irrespective of a great deal of success and inspired discoveries in brain research (starting the revelation of spot as well as composition of cells for brain mapping and stimulation of motor cortex), implantable neural probes are exposed to various obstacles that limit long-term implantation. Because of the stiff properties of the implantable probes, they often succumb to coating failure and suffer from instability of the recording/stimulating (Campbell et al. 1991). Consequently, quality of recording, stimulation capacity, along with life span of a neural probe come down to its ability to oppose or defeat increase in electrical impedance. Owing to the invasive nature of the implantable brain probe, it introduces both constant (chronic) and intense (acute) damage to tissue, however, current spotlights on the tissue/probe biointegration and life expectancy rather than the influence on the brain function.

Material defects, imperfect coating, plus unwanted mechanical stresses lead to delamination and cracking of the implantable neural devices (Prodanov and Delbeke 2016). Encapsulation or coating failure, which takes place around a week or a month after surgical implantation, can uncover the metallic interconnects to tissue. On the other hand, corrosion because of chemical degradation of the electrode material causes a couple of negative impacts. Firstly, it ruins the conductive characteristics of the electrodes as well as metals and secondly, attacks the brain with noxious chemical ingredients, subsequently, compromising the brain immune response and introduces cell death (Rivnay et al. 2017). As a result, discreet materials selection and/or synthesis are crucial in creating a neural interface which is properly encapsulated and stable chemically. Although short-term (acute) tissue reaction because of the implantable neural device in the brain can be efficiently overcome, the chronic or long-term tissue response, along with subsequent neuroinflammation at the implant location, damages the implant/tissue biointegration (Prochazka 2017). This long-term damage, which is accountable for pushing the neuron away from the implant location may be ruled by several factors. A summary of the numerous failures after the surgical implantation of a neural device into the brain is depicted in Figure 4.4.

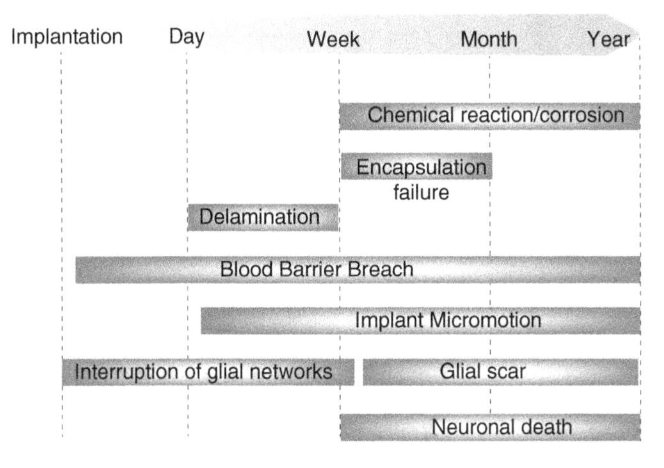

Figure 4.4 Failure Modes of Implantable Neural Devices and Duration.

Biocompatibility of the implantable neural probe can be assessed quantitatively by deriving the approximation of the population of neurons as a function of the implant distance. In addition, the extent of the neuroinflammatory response (Patrick et al. 2011) as well as the mechanisms of implantable neural probe failure can be analyzed by either optical analysis or electrochemical impedance spectroscopy. To improve implant/tissue biointegration, several methodologies have been proposed. For example, improved implant encapsulations can prevent the electrical influence from forming scars, and a reduction in implant corrosion can restrict neuroinflammation (Patrick et al. 2011). Nevertheless, mechanical inconsistency between neural interface and tissue impacts the chronic biocompatibility of the implanted neural probe. The insufficiency in mechanical conformity between the implanted device and brain together with implant micromotions are linked with scar formation (Rivnay et al. 2017; Chen et al. 2017; Das et al. 2020). A mechanically pliable device to the bran tissue is anticipated to allow the implanted probe to carry on the natural motions of the brain. Two physical quantities, Young's modulus and bending stiffness, are generally applied to illustrate rigidity or the preservation from deformation or twists. Figure 4.5 depicts the ranges of Young's moduli of various neural implants and tissue. As a consequence, the reduction in mechanical inconsistency can be tackled in two ways. First, devising a polymer-based flexible brain implant since polymers such as polyimide, parylene C, polyimide, or SU-8 are softer than metals and bulk Si. Recently, S. Guan et al. proposed the Neurotassel probe made from polyimide providing flexibility and high-aspect ratio microelectrode filaments, which can accommodate recordings up to 3-6 weeks after surgical implantation (Guan et al. 2019). Nonetheless, neural devices based on polymers are still notably stiffer than neural tissue, as shown in Figure 4.5

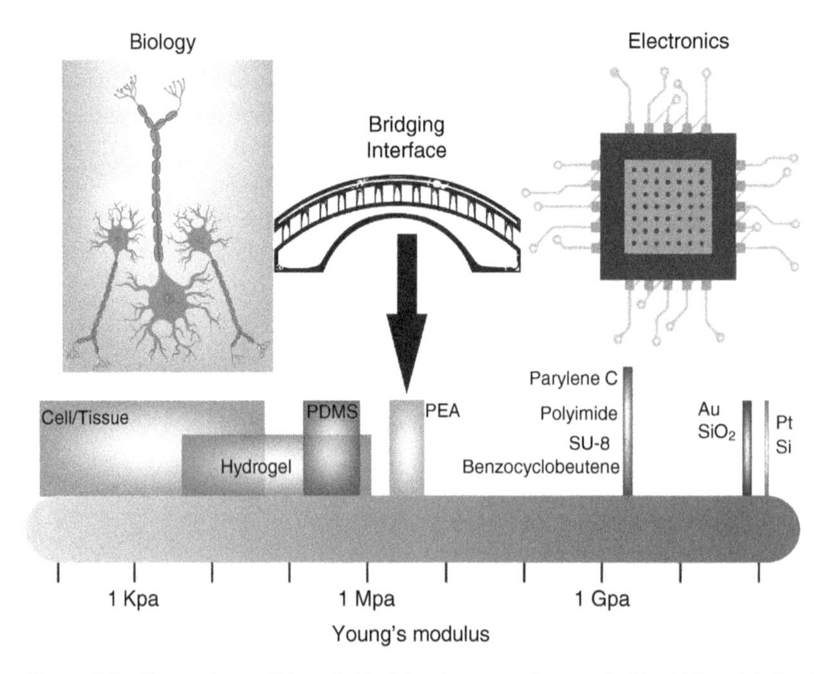

Figure 4.5 Comparison of Young's Modulus between Commonly Used Materials for Implantable Brain Device and Tissue

Shifting to increasingly agreeable materials in terms of Young's modulus, for example, polydimethylsiloxane (PDMS) or hydrogel encapsulation closes this gap (Rivnay et al. 2017). Brain implantable devices derived from the elastomeric material (e.g., PDMS) are stretchable as well as empowering chronic multimodal neuromodulation applications (Minev et al. 2015). Furthermore, standard polymer (e.g., parylene C) coating for neural implants typically yields a coating thickness in the range of a micron. For sub-micron size coating, it is also necessary to introduce new materials for brain implants to investigate the development of a plasma-polymerization strategy for coating of neural implants with thin layers (nanometer range) of poly (ethyl acrylate) (PEA). PEA is a polymer with a low hydrophilicity, elastomeric at body temperature. Its Young modulus is close to those of soft biological tissues. This polymer material has been actively used for regeneration of critical-sized bone defects (Cheng et al. 2019). On the other hand, the second technique suggests that for mechanical consistency stiff materials such as metals, polymers, and semiconductors can be applied if the characteristics dimensions of the implants are very small (i.e., in subcellular scale) (Rivnay et al. 2017). The long-term functioning of the neural probes mostly relies upon their stability, the dimensions of their materials and functionalities, adequate coatings, phyical and mechanical properties to mitigate the neuroinflammatory response. Formation of Glial scar and tissue displacement can be minimized by reducing the size of the implant, which can be realized through minimizing the cross-sectional area to lessen the stiffness of the implant. As a consequence, scaling down the neural probe geometry to below several microns improves the bendability of the implant (Luan et al. 2017). The use of innovative materials such as carbon fibre for chronic recording of neuronal activities has resulted in minor neuron loss and gliosis (Kozai et al. 2012). But producing arrays of carbon fibre is not cost-effective as it requires intensive labor and is largely operated manually. In addition, carbon fibre array configuration is limited by one electrode site per fibre. In contrast, arrays of mesh electronics arrangement known as neuron-like electronics (NeuE) have been introduced with feature size matches with the neuron axon and accommodate significantly lower bending forces, and result in less inflammation as well as promoting probe-neuron interaction (Yang et al. 2019). As a consequence, such mesh electronics can facilitate neural recording up to 3 to 8 months as compared to 7 days to 1 month with most of the implants (Guan et al. 2019). Notwithstanding, the trade-off exists since a syringe is necessary to inject the mesh electronics and lacks the accurate manoeuver over the implantation. Additionally, neuron-scale mesh electronic devices have yet to be tested on larger animals to demonstrate their chronic applicability.

4.4 MRI Compatibility of the Neural Devices

MRI is another imaging modality aiming for detection and mandatory postoperative monitoring of patients suffering from neurological diseases (Logothetis et al. 2001). In spite of that, interactions between the MRI and neurological implants constitute immense health risks to the patient, as illustrated in Figure 4.6. RF-induced heating at the tip of the implant can be significant if a patient with an medical implant is exposed to the strong radio frequency (RF) coil of an MRI system (Baker et al. 2004; Frenden et al. 2014). However, patients with brain implants more often undergo MRI tests, and injurious

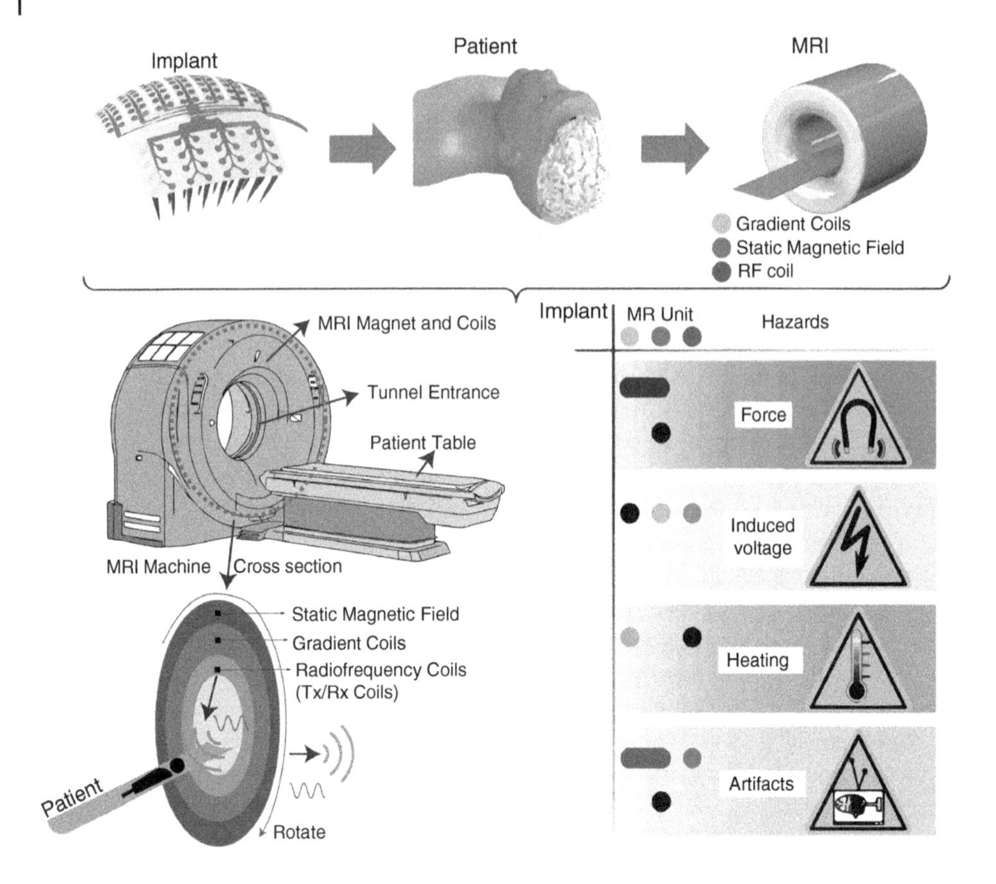

Figure 4.6 The Complex MR Environment Can Interact with an Implant in Many Different Ways Which May Cause Hazardous Conditions. Source: Based on Johannes et al. (2018).

occurrence are seldom reported. Nevertheless, this does not demonstrate that the safety of the MRI procedures. Approximately 300,000 cochlear implant patients are denied MRI scans, unless otherwise specified (Johannes et al. 2018). For around 75,000 deep brain stimulation (DBS) patients quite the opposite holds true as MRI is regarded a necessary piece of the surgical implantation, and some medical centers intentionally exceed safety regulations, which they claim are critically impractical (Johannes et al. 2018). As we are in the process of developing the next generation of brain implants, we refer to the educational standards deep brain stimulation (DBS) implants, with the aim of learning lessons from the knowledge of these ample commercial brain devices in the environment of MRI.

Currently, MRI scanners that are used for clinical purposes work at 1.5 T, with 3 T MRI systems being the second most common, whilst the permitted limit for clinical MRI imaging has recently been elevated from 7 T to 8 T for higher resolution MRI imaging (CR4 2015). In combination with progress in developing strong MRI systems as well as increases in MRI examinations, the cases of adverse issues along with implantable medical devices and MRI have become more common (Zrinzo et al. 2011). In general, materials selection for fabricating the implant is limited by longevity and biocompatibility requirements. Likewise, more restrictions in choice of materials have been applied to develop a MRI-compatible neural implant. Magnetic susceptibility is one important property of the material, which signifies the amount of magnetisation of a material when a magnetic field is applied externally. The magnetic susceptibility value acts as the amplitude of force and size of the imaging artefact that are to be anticipated in MRI from a particular material (Johannes et al. 2018).

On the other hand, an electrically conducting medical lead in the presence of RF magnetic fields in MRI scanner may concentrate the RF energy, which causes heating of tissue at the end of an implant lead. If a long conductive medical implant is placed in a strong MRI RF field, it combines with the field, and the incident field is scattered at the tip. Consequently, the implant operates like a transmitting antenna. As a result, the mismatch in impedance between the tissue and the lead induces a charge accumulation creating a current, which subsequently introduces a scattered electric field at the implant lead tip. Due to ohmic losses in the tissues, the scattered electric energy is converted into heat (Das and Yoo 2013, 2017). Furthermore, eddy currents and the induction of voltages in a medical implant can give rise to unwanted neural tissue stimulation. Implant components heating can cause burns in tissue, which may lead to tissue necrosis or permanent brain dysfunctions (Henderson et al. 2005). Moreover, artefacts in neural recordings or implant damage may occur, which subsequently make device re-implantation necessary.

Additionally, magnetic forces are generated because of the interaction of a magnetic moment with a strong static magnetic field, which can cause motion with respect to the surrounding tissue. The static magnetic field (or permanent magnet) is responsible for the inducing magnetic moment. Magnetization of any material can also cause magnetic moment if placed to an external magnetic field similar to the environment of the MRI, with an amplitude corresponding to the magnetic susceptibility as well as its volume. Consequently, eddy currents around a closed loop can also generate magnetic moment. This can happen if the patient is moved in or out of the MRI system, or changes position while imaging. A linear force is initiated by the interplay between a magnetic field gradient with a magnetic moment. The application of force onto a medical implant can create mechanical stress on the tissue as well as extreme pain (Johannes et al. 2018). The displacement of a neural implant can limit the implant functionality and requires surgical intervention.

Image noises which have no correspondent in the real object are known as imaging artefacts. There are numerous types of imaging artefacts that occur during MRI imaging and can source signal voids such as dark patches, or false appearances owing to the superposition of different voxels. Susceptibility-based artefacts are common in implants, which originate due to materials magnetization under the influence of a static magnetic field (Johannes et al. 2018). A locally alternating magnetic field occurs due to the magnetization which superimposes onto the field of the MRI static magnet. The total field deformations depend mostly on the geometry of the object and are correlated to the local mismatch in susceptibility, which causes a spatial shift in the Larmor frequency, generating inaccuracy in the imaging approximations. Additionally, significant distortions in field completely move the Larmor frequency outside the desired RF excitation frequency band, and because of the inconsistency in resonance frequency, no meaningful imaging signal is obtained. The non-homogeneous distortions also create an alteration of the Larmor frequency inside a voxel in such a way that the received resonance signal from this voxel is subdued by obstructive interference, generating dark artefacts surrounding the object. These phenomena are specifically applicable near small objects and sharp edges. Such shifts in frequency additionally produce inaccurate spatial encoding, that is, geometric distortion of the image. An artefact-free, MRI compatible implantable probe is a key for associating anatomical/functional MRI mapping and electrophysiology. Recent results (Zhao et al. 2016; Linlin et al. 2019) demonstrate that the instead of using traditional implant materials (e.g., Copper, platinum or gold) for neural electrodes, graphene encapsulated copper microwires (Zhao et al. 2016) or soft Carbon Nano Tube (CNT) fibre electrodes are able to detect reliably high-resolution individual-unit brain signals for atleast 4 months (Linlin et al. 2019). Though not as soft as the modern ultrathin-polymer-based implantable probes, and although the long-term recording period was not long, numerous properties of the CNT fibre based implants make them distinctive and beneficial for neuroscience research.

4.5 Conclusion

The progress discussed in this review indicate prompt opportunities in various branches of neural implant and engineering, principally in research, consumer, and clinical domains. These developments result directly from the distinctive capability of these technologies to intimately incorporate with the curvilinear, soft tissues of the body in ways that are not possible with traditional, wafer-based forms of electronics. Though the outcomes accomplished so far depend on diverse, physical modes of interaction with tissue, future neural devices might also integrate MRI compatibility with long-term neural modulation capability for which new sensors as well as microfluidic and microelectromechanical components in stretchable/flexible formats will be necessary. New ideas in mx device engineering and materials will be essential, as will be better insight of the physical chemistry of the biotic/abiotic interface. Further sectors for research and development of components are for scavenging of thermal, mechanical, or chemical energy and wireless communications systems. These engineering objectives, the basic science that promote them, and their connection to advances in human well-being will propel interest in the neuroscience field for many years to come.

References

I.e.c. (iec) (2015). iec 60601-2-33:2010/cor:2016 medical electrical equipment.

Baker, K.B., et al. (2004). Evaluation of specific absorption rate as a dosimeter of mri–related implant heating, *Journal of Magnetic Resonance Imaging*, 20, p. 315–320

Balasubramaniam, S., S.A. Wirdatmadja, M.T. Barros, Y. Koucheryavy, M. Stachowiak, and J.M. Jornet. (2018). Wireless communications for optogenetics-based brain stimulation: present technology and future challenges, *IEEE Communication Magazine*, 56, p. 218–224.

Birbaumer, N. (2006). Brain-computer-interface research: coming of age, *Clinical Neurophysiology*, 117, p. 479–483.

P. K. Campbell, P.K., K.E. Jones, R.J. Huber, K.W. Horch, and R.A. Normann. (1991). A silicon-based, three-dimensional neural interface: manufacturing processes for an intracortical electrode array, *IEEE Transactions on Biomedical Engineering*, 38, p. 758–768.

Checa, G.C., K. Uke, L. Sohail, R. Das, and H. Heidari. (2019). Flexible wirelessly powered implantable device, 6th IEEE International Conference on Electronics Circuits and Systems (ICECS 2019), Genoa, Italy.

Chen, R., A. Canales, and P. Anikeeva. (2017). Neural recording and modulation technologies, *Nature Review Mater*, 2.

Cheng, Z.A., (2019). Nanoscale coatings for ultralow dose bmp-2-driven regeneration of critical-sized bone defects, *Advanced Science*, 6.

Chou, K.L., (2012). Deep brain stimulation: A new life for people with Parkinson's, dystonia and essential tremor, *Demos Health, New York*.

Das, R., and H. Heidari. (2019). A self-tracked high-dielectric wireless power transfer system for neural implants, *26th* IEEE *International Conference on Electronics Circuits and Systems (ICECS 2019)*, Genoa, Italy.

Das, R., and H. Yoo. (2013). Innovative design of implanted medical lead to reduce mri-induced scattered electric fields. *Electronics letters*, 49, p. 323.

Das, R., and H. Yoo. (2017). Rf heating study of a new medical implant lead for 1.5 t, 3 t, and 7 t mri systems, *IEEE Transactions on Electromagnetic Compatibility*, 59, p. 360–366.

Das, R., F. Moradi, and H. Heidari. (2020). Biointegrated and wirelessly powered implantable brain devices: a review, *IEEE Transactions on Biomedical Circuits and Systems*, 14, p. 343–358.

Deisseroth, K. (2011). Optogenetics, *Nature Methods*, 8, p. 26–29.

Frenden, J.K.J., B. Hakansson, S. Reinfeldt, H. Taghavi, and M. Eeg-Olofsson. (2014). Mri induced torque and demagnetization in retention magnets for a bone conduction implant, *IEEE Transactions on Biomedical Engineering*, 61, p. 1887–1893.

Guan, S., et al. (2019). Elastocapillary self-assembled neurotassels for stable neural activity recordings, *Science Advances*, 5.

Gustavsson, A., (2011). Cost of disorders of the brain in Europe 2010, *European Neuropsychopharmacology*, 21, p. 718–779.

Gutruf, P., and J.A. Rogers. (2018). Implantable, wireless device platforms for neuroscience research, *Current Opinion in Neurobiology*, 50, p. 42–49.

Henderson, J.M., J. Tkach, M. Phillips, K. Baker, F.G. Shellock, and A.R. Rezai. (2005). Permanent neurological deficit related to magnetic resonance imaging in a patient with implanted deep brain stimulation electrodes for parkinson's disease: case report, *Neurosurgery*, 57.

Hong, G.S., and C.M. Lieber. (2018). Novel electrode technologies for neural recordings, *Nature Review Neuroscience*, 20, p. 330–345.

Hwang, S.W., et al. (2012). A physically transient form of silicon electronics, *Science*, 337, p. 1640–1644.

Jeong, J.W., G. Shin, S.I. Park, K.J. Yu, L.Z. Xu, and J.A. Rogers. (2015). Soft materials in neuroengineering for hard problems in neuroscience, *Neuron*, 337, p. 175–186.

Johannes, B.E., (2018). Should patients with brain implants undergo mri? *Journal of Neural Engineering*, 15, p. 041002 (26pp).

Jun, J.J., et al. (2017). Fully integrated silicon probes for high-density recording of neural activity, *Nature*, 551, p. 232–236.

Kozai, T.D.Y., (2012). Ultrasmall implantable composite microelectrodes with bioactive surfaces for chronic neural interfaces, *Nature Materials*, 11.

Kutas, M., and K.D. Federmeier. (2000). Electrophysiology reveals semantic memory use in language comprehension, *Trends in Cognitive Sciences*, 4, p. 463–470.

Lee, M., H.J. Shim, C. Choi, and D.H. Kim. (2019). Soft High-Resolution Neural Interfacing Probes: Materials and Design Approaches, volume 19.

Linlin, L., et al. (2019). Soft and mri compatible neural electrodes from carbon nanotube fibers, *Nano Letters*, 19, p. 1577–1586

Logothetis, N.K., (2001). Neurophysiological investigation of the basis of the fmri signal, *Nature*, 412, p. 150–157.

Luan, L., et al. (2017). Ultraflexible nanoelectronic probes form reliable, glial scar-free neural integration, *Science Advances*, 3, p. e1601966.

Minev, I.R., et al. (2015). Biomaterials. electronic dura mater for long-term multimodal neural interfaces, *Science*, 347.

Nabaei, V., G. Panuccio, and H. Heidari. (2020). Neural microprobe device modelling for implant micromotions failure mitigation, *2020 IEEE International Symposium on Circuits and Systems, Seville, Spain*, 2020.

Patrick, E., M.E. Orazem, J.C. Sanchez, and T. Nishida. (2011). Corrosion of tungsten microelectrodes used in neural recording applications, *Journal of Neuroscience Methodology*, 198, p. 158–171.

Prochazka, A. (2017). Neurophysiology and neural engineering: a review, *Journal of Neurophysiology*, 118, p. 1292–1309.

Proctor, C.M., et al. (2018). Electrophoretic drug delivery for seizure control, *Science Advances*, 4, p.eaau1291(1–8).

Prodanov, D., and J. Delbeke. (2016). Mechanical and biological interactions of implants with the brain and their impact on implant design, *Frontiers in Neuroscience*, 10.

Rios, G., E.V. Lubenov, D. Chi, M.L. Roukes, and A.G. Siapa. (2016). Nanofabricated neural probes for dense 3-d recordings of brain activity, *Nano Letters*, 16, p. 6857–6862.

Rivnay, J., H. L. Wang, L. Fenno, K. Deisseroth, and G.G. Malliaras. (2017). *Next generation probes, particles, and proteins for neural interfacing*. Science Advances.

Seymour, J.P., F. Wu, K.D. Wise, and E. Yoon. (2017). State-of-the-art mems and microsystem tools for brain research, *Microsystems & Nanoengineering*, 3, p. 16066.

Shin, H., (2019). Multifunctional multi-shank neural probe for investigating and modulating long-range neural circuits in vivo, *Nature Communication*, 10.

Smith, S.J.M. (2005). Eeg in the diagnosis, classification, and management of patients with epilepsy, *Journal of Neurology, Neurosurgery and Psychiatry*, 76, p. 2–7.

Swann, N.C., et al. (2018). Adaptive deep brain stimulation for parkinson's disease using motor cortex sensing, *Journal of Neural Engineering*, 15, p. 046006.

Terry, R. (2009). Vagus nerve stimulation: a proven therapy for treatment of epilepsy strives to improve efficacy and expand applications, *IEEE Annual International Conference on Engineering in Medicine and Biology Society (EMBC)*.

Wang, B., P. Yang, Y. Ding, H. Qi, Q. Gao, and C. Zhang. (2019). Improvement of the biocompatibility and potential stability of chronically implanted electrodes incorporating coating cell membranes, *ACS Applied Materials and Interfaces*, 11, p. 8807–8817.

Yang, X., et al. (2019). Bioinspired neuron-like electronics, *Nature Materials*, 18.

Yuan, M., J. Zhao, R. Das, R. Ghannam, Q. H. Abbasi, M. Assaad, and H. Heidari. (2020). Magnetic resonance-based wireless power transfer for implantable biomedical microelectronics devices, (2020). *2019 IEEE International Symposium on Signal Processing and Information Technology (ISSPIT)*, Ajman, United Arab Emirates.

Zhao, J., R. Ghannam, M.K. Law, M.A. Imran, and H. Heidari. (2019a). Photovoltaic power harvesting technologies in biomedical implantable devices considering the optimal location, *IEEE Journal of Electromagnetics, RF and Microwaves in Medicine and Biology*.

J. Zhao, R. Ghannam, M. Yuan, H. Tam, M. Imran, and H. Heidari. Design, test and optimization of inductive coupled coils for implantable biomedical devices. *Journal of Low Power Electronics*, 15:76–86, 2019b.

Zhao, S., X. Liu, Z. Xu, H. Ren, B. Deng, M. Tang, L. Lu. (2016). Graphene encapsulated copper microwires as highly mri compatible neural electrodes, *Nano Letters*, 16, p. 7731–7738.

Zrinzo, L., et al. (2011). Clinical safety of brain magnetic resonance imaging with implanted deep brain stimulation hardware: large case series and review of the literature, *World Neurosurgery*, 76, p. 164–172.

Zuo, S., D. Farina, K. Nazarpour, and H. Heidari. (2020). Miniaturized magnetic sensors for implantable magnetomyography, *Advanced Materials Technologies*.

5

Machine Learning for Decision Making in Healthcare

Ali Rizwan[1], Metin Ozturk[1], Najah Abu Ali[2], Ahmed Zoha[1], Qammer H. Abbasi[1] and M. Ali Imran[1]

[1]*James Watt School of Engineering, University of Glasgow, Scotland, UK*
[2]*Faculty of Information Technology, United Arab Emirates University, UAE*

5.1 Introduction

Machine learning and sensing technologies independently have emerged as key players for futuristic healthcare systems, characterized with P4 features (i.e., being predictive, preventive, personalized, and participatory) (Sagner et al. 2017). They are envisaged as intelligent, autonomous, and ubiquitous decision making systems for the diagnosis and treatment of diseases. The intelligence required for such decision making can be gathered by the application of machine learning on the healthcare data comprising patients' medical history, medical test reports, logs of monitoring devices, and so forth. There already exists an overwhelming amount of digital healthcare data stored in medical databases which can be exploited with the help of machine learning for developing intelligent healthcare solutions. Moreover, the use of smart sensor-based healthcare devices has increased the data generation manifolds. Such devices, including wearable fitness bands, implanted chips, auto-injectors, defibrillators, and so forth, used for monitoring and diagnosis, can generate data continuously. Machine learning can leverage from this data for the provision of seamless healthcare services in an autonomous manner (Rizwan et al. 2018). This chapter presents a very important use-case from Rizwan et al. (2020) to highlight the role of machine learning in making autonomous decisions for the provision of healthcare services. The scenario presented here involves use of the data collected for an important bio-marker, Galvanic Skin Response (GSR) measured with electrodermal activity sensors, and use of machine learning for auto diagnosis of hydration levels in the human body.

Survival of the human body depends on maintaining homeostasis, a key term in physiology with the literal meaning "same standing" (Davies 2016). It basically refers to maintaining the internal environment of a healthy human body from cellular to organ level. Some key components of the human internal environment are maintaining appropriate temperature, pressure, and chemical composition. The human body is not only required to maintain homeostasis but it is also required to perform vital functions like metabolism, respiration, digestion, reproduction, growth, exertion and so forth (Coniglio, Fioriglio, and

Engineering and Technology for Healthcare, First Edition.
Edited by Muhammad Ali Imran, Rami Ghannam and Qammer H. Abbasi.

Laganà 2020). A key ingredient required for maintaining the internal environment and performing the vital functions in humans is water. It is the major component of human body contributing about 63% to the the total body weight and 90% to blood plasma. It is also a well established fact that both dehydration and over-hydration both can disturb the equilibrium of the human body which can result in mild to severe medical implications (Aristotelous et al. 2019; Merhej 2019; Cortés-Vicente et al. 2019). On one end, dehydration can effect human performance and cause minor issues like headache or fatigue. On the other end, it can lead to severe metabolic, gastrointestinal, circulatory, or urological complications (Manz 2007). Similarly over-hydration can cause medical complications like oedema (Evans, Maughan, and Shirreffs 2017). Apart from the risk of medical complications, dehydration or over-hydration can also affect the regular lifestyle of human beings. Studies show that maintaining appropriate hydration levels is even more important for sick children and the elderly as dehydration can cause additional complications for them (Cortés-Vicente et al. 2019; Pross 2017). It is, therefore, crucial to maintain an appropriate hydration level in the human body.

Hence, there is a need for hydration monitoring or detection tools for the assessment of the hydration level in the human body, so that the measures required can be taken in case of dehydration or over-hydration. Recent studies show much interest in hydration monitoring tools, but most of them rely on manual entry by the user, for the water intake, in a mobile app. Some other recent studies are introducing bottle mount recorders to log water intake (Howell et al. 2019; Lee et al. 2015). None of these solutions assess the actual hydration level in human body; instead they just maintain a log of water intake. However, there is a dire need for a non-invasive wearable type of hydration monitoring device. A key challenge in the development of such a device is the identification of a bio-marker for which data can be recorded in a non-invasive manner, and it should indicate the hydration level with promising accuracy.

Some commonly discussed signs or parameters in the literature for the hydration level detection include Dry Mucous Membrane, dry axilla, tachycardia, poor skin turgor, urine color, urine specific gravity, low systolic blood pressure, blood urea nitrogen to creatinine ratio, total body water, saliva flow rate, saliva osmolality, plasma or serum osmolality, and bio-electrical impedance (Fortes et al. 2015; Armstrong 2005; Garrett et al. 2017; Armstrong 2007). There are bottlenecks associated with the use of each of these parameters or bio-markers. For example, most of them involve invasive methods and require clinical arrangement for data collection, while others involve biochemical analysis of serum or fluid. Because of such limitations it is not feasible to use them in wearable non-invasive monitoring and diagnosis solutions for hydration detection.

From the parameters mentioned above, bio-electrical impedance analysis (BIA) is a non-invasive method in which a light current is passed through the body and the resistance faced by the current is measured. This resistance to current is used as a measure of fat mass and fat-free mass, and total body water (Khatun et al. 2019; Marra et al. 2019). The conventional methods of the BIA measurement are complex and require special equipment not suitable for the continuous monitoring. But BIA methodology provides a proof of concept that there exist a basic correlation between the electric resistance of the body and the water content in it. Depending on the water contents in the human body resistance can vary between few ohms and thousands of ohms. A separate study (Fish and Geddes 2009) shows that more than 99% of the resistance is faced at skin level.

Taking inspiration from BIA, the authors of this chapter have gathered data of GSR and developed a solution for the detection of hydration levels based on the GSR data. GSR is basically the measure of skin conductance, which is just the reciprocal of the skin resistance. As shown in Figure 5.1, skin conductance varies between hydrated and dehydrated states. Therefore, in a systematic way, this skin conductance data labeled as GSR data is collected from multiple participants using an EDA sensor. The EDA sensor basically used to study the sympathetic behavior in humans. But it works on the same principle of passing light current form the human body at skin level. Hence the GSR data collected from EDA sensor is then exploited with machine learning to detect the hydration level. At the end of this study, it is found that GSR data can be used as an indicator of hydration levels in human body. Major contributions of this study include:

- Identification of the right body posture for the GSR data collection for hydration level detection.
- Identification of the appropriate interval of time, window of time, for which the data sample should be gathered.
- Identification and extraction of features that encompass enough information for the identification of hydration levels.
- Development of a machine learning model for the auto-estimation of hydration levels in the human body using GSR data.

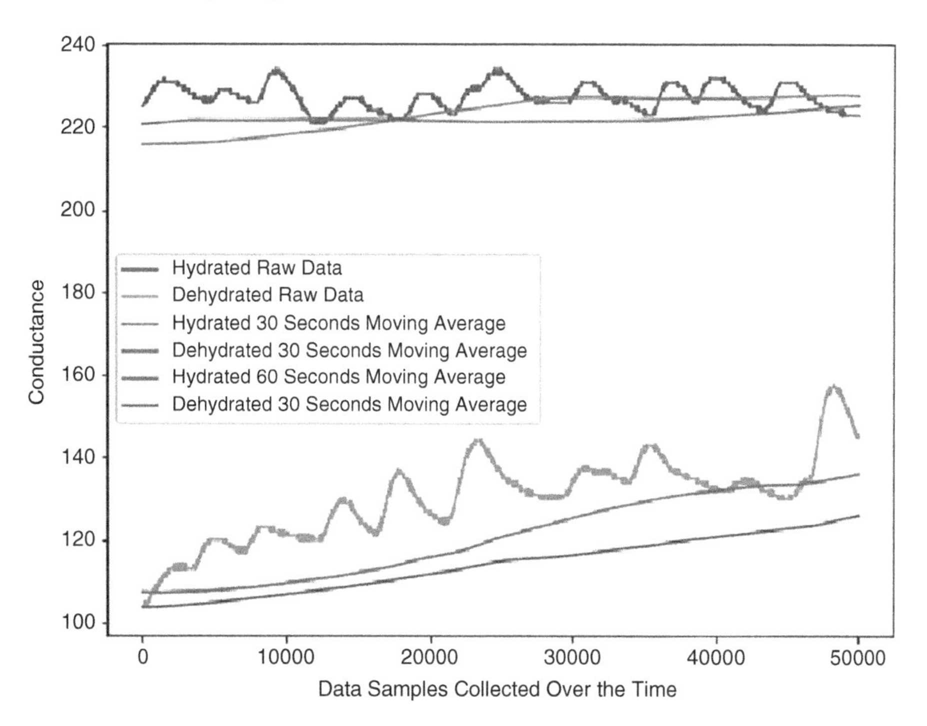

Figure 5.1 Data Collected in Hydrated and Dehydrated State in Sitting Posture with a Moving Average Representation for 30 and 60 Seconds

5.2 Data Description

The electrodermal activity aka GSR data used in this study is collected from five individuals using EDA sensor available on BITalino Kit (Da Silva et al. 2014). All the participants are between 25 and 35 years of age. One of them is female, and the rest are male. They belong to different ethnic groups and have no known medical history of edema or hypohydration. Since the hydration level in the human body is studied here as a binary problem, that is, a person is considered either hydrated or dehydrated. Therefore, the data used is also for these two states (hydrated and dehydrated). In particular, when the data is collected from the participants after fasting at least ten hours, it is labeled as dehydrated. Whereas data collected when the participant has been drinking water frequently and has had drunk water within one hour before the data collection is labelled as hydrated.

Another aspect considered here is the impact of body posture on electerodermal activity because it varies with body movements and changes in body postures. Hence, it is also important to identify the body posture, for which the EDA data is collected. This can help to achieve better accuracy for hydration level detection. In this study, therefore, hydrated and dehydrated state data is collected in two common body postures: sitting and standing. For a posture-independent study, on the other hand, hydrated and dehydrated state data collected in both postures is combined and fed to a model that is unaware of the body posture.

To summarize, there are three datasets:

1) Dataset for sitting posture
2) Dataset for standing posture
3) Dataset for posture independent scenario, it is created by combining datasets 1 and 2.

Data is collected from every participant for two hours, out of which, one hour is for the hydrated state and other one hour is for the dehydrated state. As hydrated and dehydrated state data is collected in the two body postures separately, therefore, in each state, data is collected for sitting and standing postures for a period of half an hour. The total recorded data, from all participants in the both states and body posture, is for 10 hours. The BITalino kit used here basically measures the skin's conductance level in μS and is denoted as G. It is simply the reciprocal of skin resistance calculated as follows:

$$G = \frac{1}{R}, \tag{5.1}$$

here R represents skin resistance measured in MΩ:

$$R = 1 - \frac{A}{2^n}, \tag{5.2}$$

where A is the analogue to digitally converted value of the electric signal of the data collection channel BITalino at resolution n, while n represents the number of bits used to store digital output after conversion from the analogue signal. Here data is collected using a resolution of 16 bits on the BITalino kit channel.

Lastly, for the auto diagnosis of hydration levels with machine learning algorithms, statistical features are extracted from GSR data, and those features are fed to algorithms as input instead of raw data. For this objective, raw GSR data is split into small epochs for four different length of time intervals (window) and features like mean, mod, variance, and so forth, are extracted from those epochs. The numbers of instances created for feature values for each state, posture and window of time are presented in Table 5.1.

Table 5.1 Number of Samples Generated after the Feature Extraction from the Segments of Raw Data Collected for Different Time Intervals, Hydration States, and Body Postures

Posture	State	Number of Samples for Window Size			
		30 sec	45 sec	60 sec	75 sec
Standing	**Hydrated**	290	205	150	125
	Dehydrated	290	205	150	125
Sitting	**Hydrated**	300	200	150	120
	Dehydrated	300	200	150	120
Independent	**Hydrated**	640	425	325	260
	Dehydrated	640	425	325	260

5.3 Proposed Methodology

The main steps of the methodology followed in the development of the hydration level detection model are presented in Figure 5.2 and illustrated here briefly.

5.3.1 Collection of the Data

For the fine data collection, a specific protocol needs to be followed such as placing electrodes on a specific place on the body for a certain interval of time, and low voltage current should be passed. For this objective, the data collection process adopted here adheres to

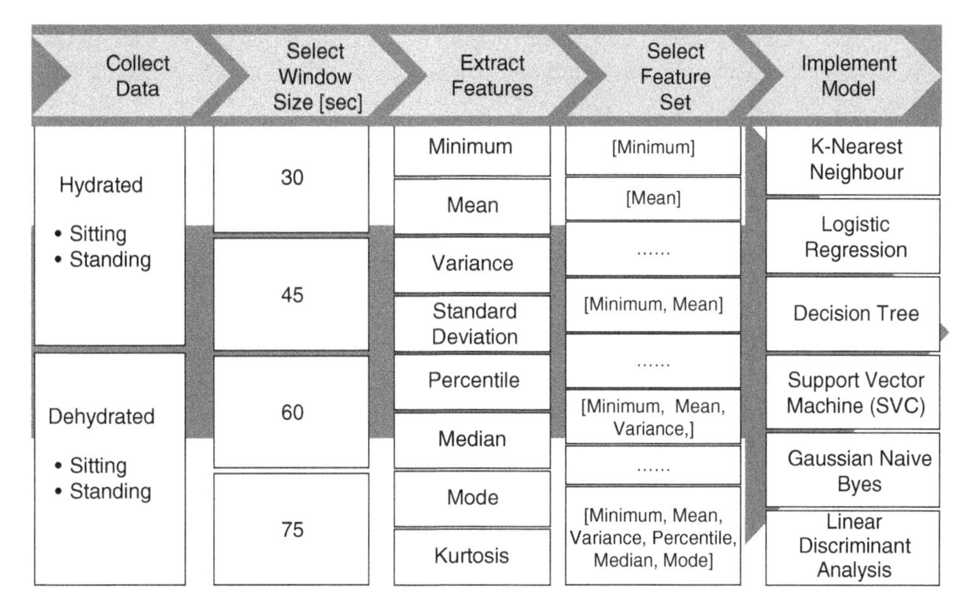

Figure 5.2 Key Steps of the Methodology Used for the Development of Hydration Level Detection Model

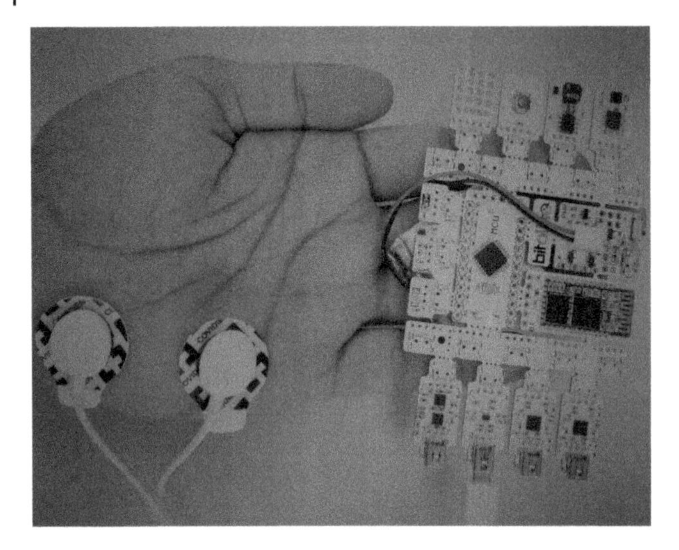

Figure 5.3 BITalino Kit Used for the Data Collection of GSR

the guidelines provided on the EDA data collection by the ad-hoc Committee of the Society for Psychophysiological Research on Electrodermal Measures (CSPREM) (Boucsein et al. 2012). In addition, approval is sought from the respective ethical committee of the University of Glasgow before the collection of data. A BITalino kit, shown in Figure 5.3, is used for the GSR data collection at a sampling rate of 1 MHz and 16 bit resolution, the highest precision options available on BiTalino kit. For the GSR data collection, pre-gelled active and passive electrodes are placed at hypothenar sites on the palm of the non-dominant hand as recommended by CSPREM. Data is recorded in samples of five to fifteen minutes for every participants to avoid any issues due to sweating on the palm. For each participant, data is collected in two postures and two states as described in Section 5.2. In total, datasets for three scenarios are analyzed. These scenarios are defined based on the body posture in which the data is collected. In one scenario data is collected in the sitting posture, and in an other scenario data is collected in standing posture and in third scenario the data collected in sitting and standing postures are combined to create another dataset, so a posture-independent models can also be developed and evaluated.

5.3.2 Selection of the Window Size

Data is initially recorded in samples of 5 to 15 minutes for every participant using BITalino kit. Then this data is split into smaller chunks of data for the different intervals of time named as windows of time, such as data samples of 30 seconds, 45 seconds, and so forth. The objective here is to identify and propose an optimal window of time for which the data collected can help to identify the hydration level with better accuracy. The statistical features used as input in the models are extracted from these data chunks of different intervals of time. In this study the following four window sizes are studied:

$$W \in \{30, 45, 60, 75\}.$$

where W presents the window of time in seconds for which the data chunk is used for feature extraction. A key advantage of identification of an optimal window size is that once a window size is identified for the best accuracy, it can also be used for the the identification of the hydration level in real time. This means that data can be collected for that length of time in real time and the model should be able to tell whether the person is hydrated or dehydrated with certainty. In this study an exclusive window approach is used instead of overlapping windows. This is done because the data is collected at very high resolution, that is, in a short interval something like 30,000 data points are gathered. So getting some data from the previous window with an overlapping operation is not expected to have significant impact. Secondly, even with an exclusive window approach, after feature extraction sufficient numbers of samples are available for model training and testing. Another reason for the use of the exclusive window operation is that the tonic behaviour of the GSR data is more important than the phasic behaviour (Boucsein et al. 2012).

5.3.3 Extraction of the Features

Once the data is split into small chunks for a specific window size, then features are extracted from those chunks of data. The feature space F used in this study comprises the following features:

> $F \in$ {Minimum, Mean, Variance, Entropy, Standard Deviation, Percentile, Median, Mode, Kurtosis}.

In this research, nine statistical features are extracted for each of the three datasets and for the each of the four window sizes. For example, a window size of 30 seconds is selected for the sitting data, followed by splitting it into non-overlapping chunks of 30 seconds and statistical features, such as mean, mode, and so forth, are calculated for each data chunk. The feature extraction function results in nine data vectors, each of which comprises the values extracted for each statistical feature from F. The number of instances or samples generated in feature vectors for each window size, body posture, and hydration state as the result of feature extraction function are shown in Table 5.1.

5.3.4 Selection of the Features

As a result of feature extraction, nine feature vectors are generated, but the identification of the right combination of features that can generate the best accuracy for the detection of hydration level is a separate important task. One possibility is to use different feature selection algorithms to select a subset of features that can be used in the final model. But every feature selection approach has its own pitfalls. Another approach is to try and evaluate all possible combinations of features. This approach is computationally costly and time consuming with a conventional sequential approach. In this study, the latter exhaustive heuristic approach is used where all the feature combinations are tried and evaluated. It could be made possible because of the use of our parallel computing approach presented in Section 5.3.7. It is a model driven feature selection approach, that is, values of all combinations of features are fed to models as input and performance of the models is evaluated, and the combination that generates best performance is selected. For a single posture-based dataset (e.g., for the sitting posture dataset) the total number of combinations of features are tried and evaluated are $(2^F - 1)$, where f is the number of features in the feature space.

5.3.5 Deployment of the Machine Learning Models

There exists a plethora of machine learning models, which are commonly grouped into two broad categories: supervised learning and unsupervised learning. Unsupervised algorithms, also known as clustering algorithms, are used where the predicate is unknown; in other words they are applied on the unlabeled data. They are commonly used to label the data and identify important features in data. Supervised learning algorithms are used where the predicate is known that is, data is already labeled. Supervised algorithms are further grouped into two main categories: i) regression algorithms used for the prediction of continuous values in data and ii) classification algorithms used for the prediction of categorical values in data. Since the predicate in this study, the hydration state, is known and it is a categorical variable, therefore, the supervised machine learning algorithms for classification are used here for the identification of hydration level based on the GSR data collected in two states. Thus, the model is trained with GSR data for two states, and when new unfamiliar GSR data is fed to the trained model, the model can classify it either as hydrated or dehydrated.

In this study, six machine learning algorithms of classification are trained and tested on data of the three scenarios to identify the hydration state. The six algorithms used are i) Logistic Regression (LR), ii) Gaussian Naive Byes Classifier (NB), iii) Support Vector Machine-Based Classifier (SVC), iv) K-Nearest Neighbour (K-NN), v) Decision Tree (DT), and vi) Linear Discriminant Analysis (LDA). All the six algorithms are trained, validated, and tested on the three datasets with all combinations of feature extracted over the four windows stated already. The number of models N_m trained and evaluated here can be computed as follows:

$$N_m = [A \times S \times W \times (2^F - 1)].$$

where A, S, W, and F are the count of algorithms, datasets of postures based scenarios, window sizes, and combinations of features used in this methodology respectively. As mentioned already, instead of raw data, data values generated after feature extraction are used for the models' training, validation and testing: 70% of the feature data is used for training with the three-fold cross-validation. The remaining 30% of the data is kept separate and used for the testing of the models.

Another important step in model development is the tuning of algorithm-specific parameters. In this regard, with a set of values for algorithm-specific parameters, models are trained and validated to identify the values of the parameters for which models consistently generate the best performance. Being consistent is also very important for models, such that they should perform well not only on one dataset but also when new datasets are fed to the model. Hence, three-fold cross-validation is also performed on the training data with the user defined values for key parameters explained in Pedregosa et al. (2011) and listed in Table 5.2. In three-fold cross-validation, training data is split into three sub-datasets. Then a model is trained and validated for all tuning parameters in such a way that in each turn one

Table 5.2 Algorithms-Specific Important Hyperparameters and Their Values Used in Model Optimization

Algorithm	Hyperparameter
KNN	Metrics: [Minkowski, Euclidean, Manhattan]
	Weights: [Uniform, Distance]
	K: x, 1 ≤ x ≤ n − 3, where n = No. of Samples
LRA	Penalty = [L1, L2]
	C: [0.001, 0.01, 0.1, 1, 10, 100]
DT	Criterion: [Gini,Entropy]
	Max depth:x, 3 ≤ x ≤ 25
SVC	Kernel: [Linear, Rbf, Poly]
	C: [0.1, 1, 10, 100, 1000]

of the sub-dataset is left out for the model validation, while the rest of the two sub-dataset are used for the model training. In three-fold cross-validation, this training and validation process is repeated three times, every time a new sub-dataset is used for the validation and rest of the two sub-datasets are used for training. Cross-validation helps to identify the model that is not only accurate but it is also persistent in its performance. It reflects that model is not biased for some parameter values and it should perform well for all new coming dataset. After that, the cross-validation models are also tested against the testing data. For more detail on cross-validation please consult Schaffer (1993). At the end, a model is selected that not only is consistent in cross-validation but also outperforms on testing data.

5.3.6 Quantitative Assessment of the Models

Another important step of machine learning model development is the evaluation of the performance of the models, for which there exist a range of metrics in the literature (Hossin and Sulaiman 2015). The selection of the right set of metrics depends on the nature of the problem and the choice of algorithms. Since it is a healthcare problem studied as a binary classification problem, some of the most relevant metrics commonly used for such healthcare problems are applied for the performance evaluation of the models here and are listed below:

$$TPR = \left(\frac{TP}{TP + FN}\right), \tag{5.3a}$$

$$TNR = \left(\frac{TN}{TN + FP}\right), \tag{5.3b}$$

$$CCR = \left(\frac{TP + TN}{P + N}\right), \tag{5.3c}$$

$$\gamma = \left(\frac{TP}{TP + FP}\right), \tag{5.3d}$$

where TPR is the true positive rate, also known as sensitivity, and represents the detection rate of the correct number of dehydrated cases, labeled as true positive cases *TP*, with respect to the sum of true positive, and false negative cases *FN*. False negative cases are actually dehydrated cases, but they are detected as hydrated cases. Similarly true negative cases *TN* are the hydrated cases detected correctly and false positive cases *FP* are the hydrated cases but they are detected as dehydrated ones, whereas TNR is the true negative rate, also called specificity, and it is the measure of the detection of correct number of hydrated cases *TN*, with respect to the sum of the true negative *TN*, and false positive cases *FP*. Here, the overall accuracy (%), also called the correct classification rate (CCR) is initially used to assess the performance of all the algorithms in all possible scenarios (i.e., for all window sizes, postures, and combination of features). When the best performing models are shortlisted based on CCR score in different posture-based scenarios, metrics like TPR, TNR, and γ (the precision also called recall), are used to assess the detailed performance of the shortlisted best model for each body posture based scenario.

5.3.7 Parallel Processing

As already mentioned and elaborated in the steps of methodology listed above, the adopted model development approach involves training, validation, and testing of six machine learning algorithms on $(2^F - 1)$ combinations of features, extracted over four window sizes from three datasets. This whole process is very time consuming and computationally costly if performed in a sequential way, and to address this challenge parallel processing is applied in two stages. In the first stage, features are extracted over four window sizes for three datasets. In the second stage all six algorithms are trained, validated, and tested for all the combinations of features in parallel. The parallel approach employed in this study is presented as Algorithm 1. At the first step of parallel processing, each of the three datasets (D) is split into small segments (smaller datasets) for the four windows of time (W), and then features (F) are extracted from each segment. Twelve new feature datasets (D_f)

Algorithm 1: Parallel Processing Approach for the Development of Hydration Level Detection Model

---

1 **Datasets (D)** $\in$ {sitting (D_{sit}), standing (D_{sd}), independent (D_{ind})}
2 **Features (F)** $\in$ {minimum, maximum, variance, entropy, standard deviation, percentile, median, mode, kurtosis}
3 **Window Sizes (W)** $\in$ {30, 45, 60, 75} seconds
4 **Algorithms (A)** $\in$ { K-NN, LR, DT, SVC, NB, LDA}
5 **do in parallel**
6   Extract F from each dataset in D for each window size in W
7   **return** *the data extracted in Step 6*
8 Create holistic feature datasets (D_f);
9 **do in parallel**
10   Train a model for each algorithm in A with cross-validation on D_f
11   Test the models developed in Step 10
12   **return** *the best models for each dataset in D*

---

are generated comprising the feature vectors extracted over the four window sizes from the three datasets. At the second step, six machine learning algorithms are trained and evaluated for each combination of features in feature datasets.

5.4 Results

The best results, in terms of CCR, produced by different algorithms are presented in Table 5.3. These are the results generated after the optimization of algorithms over the hyper-parameter space listed in Table 5.2 and exploitation of feature space listed in Section 5.3.3. From Table 5.3 it can be observed that classical machine learning algorithms can identify hydration levels with accuracy of up to 87.78%. These results are promising for hydration-level detection in the human body based on GSR data that can be collected with a non-invasive approach. A comparison of the performance of different classification algorithms, measured with CCR, is presented in Figure 5.4. It can be seen that K-NN overall outperforms the other algorithms for the posture-specific

Table 5.3 Performance Comparison of Machine Learning Algorithms for Different Window Sizes and Posture-Based Scenarios

Model	Data set	Accuracy (%) for the window size:			
		30 (sec)	45 (sec)	60 (sec)	75 (sec)
KNN	Sitting	76.67	82.50	**87.78**	86.11
	Standing	71.26	69.92	**83.33**	73.33
	Independent	**76.82**	74.51	69.23	73.72
LR	Sitting	68.89	79.17	80.00	79.17
	Standing	60.92	73.17	74.44	73.33
	Independent	69.53	68.63	64.62	68.59
DT	Sitting	76.67	80.83	81.11	84.72
	Standing	71.26	70.73	76.67	74.67
	Independent	72.14	72.94	69.23	74.36
SVC	Sitting	74.44	76.67	81.11	80.56
	Standing	74.14	71.54	80.00	76.00
	Independent	75.52	75.69	68.21	73.72
NB	Sitting	68.89	75.00	76.67	76.39
	Standing	66.67	74.80	73.33	73.33
	Independent	73.18	69.41	65.13	68.59
LDA	Sitting	69.44	77.50	75.56	77.78
	Standing	62.64	71.54	74.44	70.67
	Independent	71.09	68.63	64.62	67.31

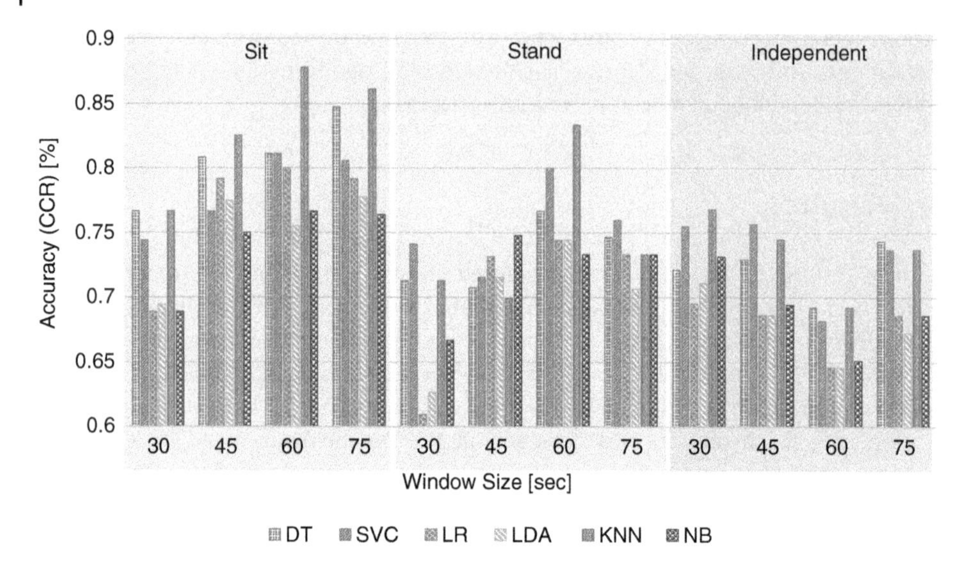

Figure 5.4 Performance Comparison of Different Algorithms in Different Body Postures at Different Window Sizes

Table 5.4 Performance of the Best Performing K-NN Classifier against Different Metrics in Each Posture-Specific Scenario

Posture	Window size [sec]	Precision	TPR	Specificity	CCR
Sit	60	0.81	0.95	0.82	0.88
Stand	60	0.76	0.93	0.76	0.83
Independent	30	0.71	0.84	0.71	0.77

and posture-independent scenarios. For the feature combinations A, B, and C presented in Table 5.6, K-NN gives an accuracy of 87.78% for the hydration level detection in the sitting posture scenario. On the other hand, for the standing and posture independent scenarios, the K-NN produces an accuracy of 83.33% and 76.82% for feature combinations D and E, respectively. K-NN yields these results for the sets of tuning parameters as follows: 1) for sitting posture, metric: *Minkowski*, $K = 15$, weights: *uniform*; 2) for standing posture, metric: *Manhattan*, $K = 10$, weights: *uniform*; 3) for posture independent scenario, metric: *Manhattan*, $K = 60$, weights: *distance*. Once the best performing models are found for all three posture-based scenarios, then their performance is evaluated against the advanced evaluation metrics presented in Table 5.4. Since K-NN exhibits the best performance in all three scenarios, detailed analysis of its performance against these advance performance metrics is presented in Table 5.4. It can be seen that the highest precision, sensitivity (TPR), specificity, and overall accuracy are achieved in the sitting scenarios, which are 0.81, 0.95, 0.82, and 0.88, respectively. From Figure 5.5 it can be observed that the best performing K-NN based model, with the parameters mentioned above, exhibits TPR of 0.95 and TNR of 0.82 in the sitting posture scenario. This means that the model could identify 95% of the

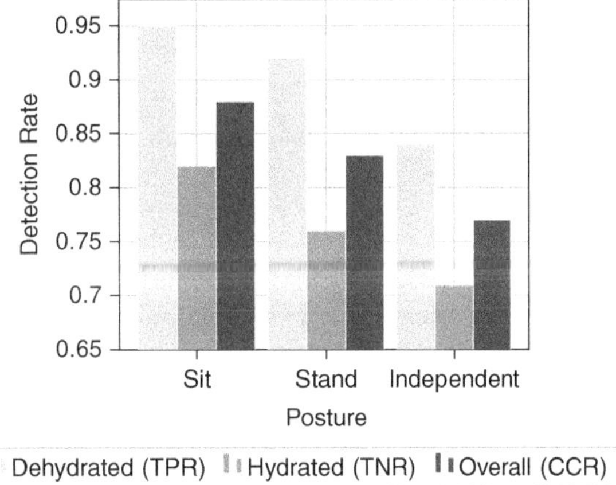

Figure 5.5 Class-Specific Performance of the K-NN Based Best Models in Each Posture-Specific Scenario

dehydrated state instances and 82% of the hydrated state instances correctly. Similarly, for the standing data, a TPR of 0.92 and a TNR of 0.76 are achieved, whereas for the data of posture independent scenarios, 0.84 TPR and 0.71 TNR are achieved. It can be observed that TPR of dehydrated state detection is even higher than the overall accuracy.

The confusion matrix in Table 5.5 also supports this observation, where the values in red boxes show the false negatives and false positives whereas the values in green represent true positives and true negatives, also support this observation. It can be seen that there are fewer examples of dehydration instances, that is, 2 of sitting, 3 of standing, and 29 of independent scenario, confused and wrongly detected as hydration states as compared to hydration instances, i.e., 9, 12, and 60 instances of sitting, standing, and independent scenarios respectively, confused as dehydration instances. During data collection for dehydrated states, it is observed that GSR is low overall (i.e., skin conductance is comparatively low because of lower water contents) and comparatively steady. That steady behaviour and fewer variations can be a reason for more accuracy for the detection of dehydrated cases as compared

Table 5.5 Class-Specific Empirical Results via Confusion Matrix for the Best-Performing K-NN Model in Three Different Scenarios: Sitting, Standing, and Independent

			Predicted	
			Dehydrated	**Hydrated**
		Dehydrated	38	2
SITTING	**Actual**	Hydrated	9	41
		Dehydrated	37	3
STANDING	**Actual**	Hydrated	12	38
		Dehydrated	149	29
INDEPENDENT	**Actual**	Hydrated	60	146

Source: Rizwan, A., Ali, N.A., Zoha, A., Ozturk, M., Alomaniy, A., Imran, M.A., & Abbasi, Q.H. (2020) (© [2020] IEEE).

to hydrated ones. Hence, if a reminder or alarm system is set for dehydrated state detection, that is, the alarm goes off only when the dehydrated state is detected, then fewer false alarms can be expected.

It is also observed that algorithms commonly perform better on the window size of 60 seconds for the posture-specific scenarios. It is, therefore, safe to say that 60 seconds is an optimal window size for the feature extraction for posture-specific scenarios. Another common behavior observed for almost all algorithms, as can be seen from Figure 5.4, is that they generally perform better on posture-specific data. This reflects that GSR not only varies because of hydration level but also because of body posture.

Based on these results, it can be stated that, with the selection of appropriate window size, features set and machine learning algorithm, GSR data can be used to identify the hydration level in the human body. More specifically, it can be concluded if GSR data is collected for 60 seconds in the sitting posture from an individual, the K-NN model can detect with an accuracy of almost 95% if the individual is dehydrated.

5.5 Analysis and Discussion

Like many real data-based healthcare studies the main objectives of this study are the identification of the appropriate body posture and optimal interval of time for the data collection of bio-markers and selection of the right combination of features and reliable algorithm for the model development for the auto diagnosis. In the light of the analytical study here, the impact of these factors is discussed in this section.

5.5.1 Postures

An important aspect to be considered for the collection of data in human beings for a bio-marker is the right posture, particularly where the readings vary with the body posture as it the case here. GSR varies with body posture; therefore, the data is collected in two important body postures: sitting and standing. Models are trained and evaluated for both body postures independently. Besides that a third dataset is created by combining both datasets for a posture-independent study where the algorithm is unaware the body posture. From a first glance at Figure 5.4 it can be observed that the algorithms mostly perform better on the data collected in the sitting state. For the standing posture, performance remains better for the window size greater than 45 seconds as compared to the posture-independent scenario for most of the algorithms. On the other hand, for the posture-independent scenario, the performance of the algorithms varies a lot, not only across the window sizes but also on the same window size, making it the most unreliable type of data. For the independent data, it is observed, only for the smaller window size 30 seconds, algorithms perform better as compared to their performance over the sitting and standing data, except the DT. For overall better performance with a visible trend, sitting data seems the most reliable dataset for hydration level detection. This can be associated with the fact that the participants are more calm in sitting posture and they have fewer body movements. These factors help to reduce the overall noise in the data. Thus it can be concluded that for hydration level detection the sitting posture is better for data collection.

5.5.2 Window Sizes

Another key step in the development of an auto-diagnosis solution for healthcare services is the collection of right amount of data for a bio-marker. It is important to know how much data is good enough for a model to make a correct decision with good certainty. Similarly in this study, it is important to identify the length of time interval for which the GSR data collected can help to detect hydration level with greater certainty. For this objective, therefore, four different length of time intervals or window sizes are employed. Data is split for those window sizes, and, for each window size, models are trained and evaluated.

From Figure 5.4 it can be observed that performance of the models seems to improve with an increase in window size, particularly for sitting posture data, which is found to be the most reliable. It means that if data is collected for longer intervals hydration level can be detected with a greater accuracy. However, this trend does not continue after a certain length of time. It persists for the increase in window size from 30 seconds to 60 seconds, after which performance starts decreasing for the majority of the algorithms. Therefore, it can be inferred that just increasing the window of time longer than 60 seconds does not seem to help. The same is approved by the study of the behavior of models over the four window sizes on the standing data. For the standing data, again performance seems to improve with increase in window size until 60 seconds for the majority of the algorithms but it decreases with a further increase in the window size. Performance of the most of the models has a unique decreasing trend for the posture-independent data from window size 30 to 60 seconds, and it improves later. It looks just the opposite trend to the other two datasets.

From the discussion above it can be concluded that after the selection of a reliable body posture, like sitting, if data is collected for 60 seconds, a model can detect the hydration level with better accuracy. Decrease in accuracy after the 60 second interval is in fact helpful in deciding a cut of time; otherwise intervals even longer than 75 seconds need to be explored. A window of 60 seconds is not only good in terms of accuracy, but it is also practical to collect data for that interval of time, particularly for a real-time implementation of the model. Hence, if real-time GSR data is collected in the sitting posture for 60 seconds, the model should be able to detect the hydration level with a greater accuracy. For example, K-NN based model can detect it with an accuracy of up to 87.87%.

5.5.3 Feature Combinations

Selection of the right posture and interval of time for the data collection of a bio-marker is not enough, because another important factor that influences the performance of the model is how data is fed as input to the model. Data may be fed to the model in the raw form, such as GSR data as it is, or a set of features can be extracted from raw data and features data can be used as input for the model. Since GSR is single bio-marker data collected at high resolution and it has sudden variation because of the impulsive nature of GSR, the use of the raw data is not a practical option. To address this challenge, statistical features are extracted instead. It is easy to extract such features, and computational cost for their extraction is also low, but selection of the right set of features to be extracted and used as input is very important. A particular set of features may achieve exceptional accuracy for certain data, but it may not do the same for the other datasets. Therefore, the selection of the right combination of features is not only highly important for achieving better accuracy but it is also important to develop a consistent reliable model.

In this study, therefore, the most commonly used statistical features are extracted and models are trained and evaluated for each combination. In total ($2^F - 1$) combinations are evaluated. Each feature is evaluated individually as well to find their individual significance in the model. Some of the features with an accuracy around 50% do not perform very well. Those features include: standard deviation, variance, entropy, and kurtosis. If there is any single feature that can generate exceptional accuracy it is "mean," since it alone yields an accuracy of 82% as an input feature when extracted for the window size of 75 seconds on the sitting data and fed to K-NN model. The remaining could generate accuracy of up to 66% when used as single feature in models.

It can be seen that individual features cannot generate the best accuracy so combinations of features are used. Since K-NN has shown the best performance in this study, in this section only five combination of features with top performance on the K-NN are discussed. Those top five performing combinations are listed in Table 5.6. Their corresponding window and dataset in terms of body posture is also mentioned against them. A performance of K-NN for all these five combinations is also compared in Figure 5.6. This feature based comparison also confirms the perception developed based on the analysis in the above two sections, that is, sitting data is better in quality as it shows better results. Similarly all these combinations yield better performance for sitting data and performance improves with increase in the window size.

From Figure 5.6 it can be observed that impact of window and posture is greater than the change in the feature combinations. This means that for the same posture and window size all combinations have similar performance with little or no variation, but performance varies even for the same combination when the window size or posture is changed. It is observed that no single combination could perform consistently across all window sizes and posture-based dataset. Thus, selection of the single feature set that may be used on data of multiple body postures and intervals of time is challenging. One possible reason for the variation in the performance for the same feature combination across different posture

Table 5.6 Features Combinations with the Best Performance for Different Window Sizes and Postures

ID	Combination	Window size (sec)	Posture	Accuracy [%]
A	Mean, Variance, Entropy, Percentile, Standard Deviation,	60	Sit	87.78
B	Mean, Variance, Entropy, Percentile, Standard Deviation, Median	60	Sit	87.78
C	Mean, Entropy, Standard Deviation, Percentile	60	Sit	87.78
D	Minimum, Mean, Entropy, Mode, Standard Deviation, Kurtosis	60	Stand	83.33
E	Minimum, Mean, Variance, Entropy	30	Independent	76.82

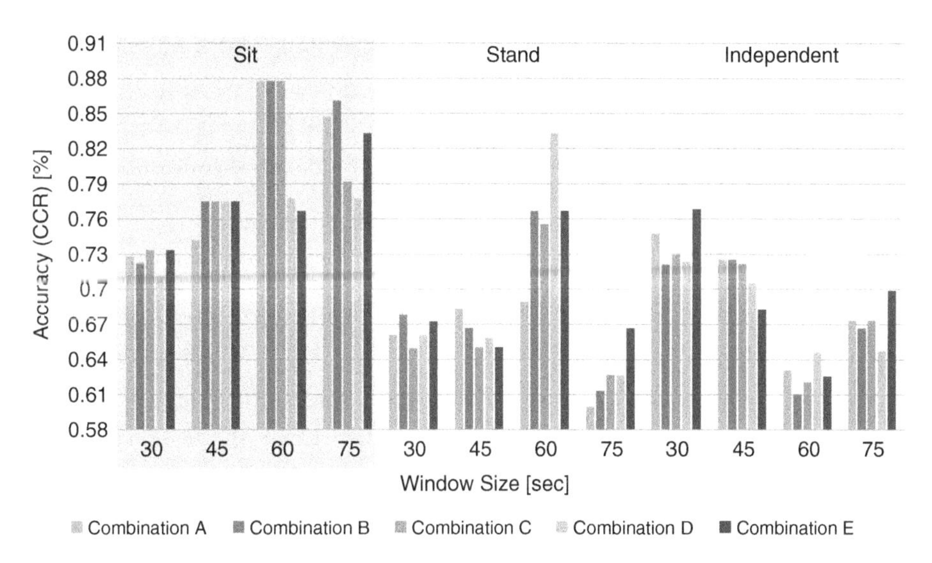

Figure 5.6 Comparison of the Performance of K-NN for a Different Combination of Features in Different Body Postures and Different Window Sizes. Please refer to Table 5.6 for the details about the combinations.

data can be the posture-specific variations in the data, where the same features may not capture those variations for different postures.

Another interesting finding of the analysis of feature combinations is that performance of the individual features is independent of their performance when they are used in combination. It means that a single low-performing feature can give better accuracy when used in combination with other features and one good-performing feature can yield bad accuracy when combined with some other features. Besides that, from Table 5.6 it can be seen that A, B, and C, the combinations of best accuracy for the sitting data, have four features in common. Moreover, it can also be observed that the three of them are also present in the Combination D, the combination that yields best accuracy for the standing posture data.

5.5.4 Machine Learning Algorithms

As already mentioned there exists a range of machine learning algorithms, and many of them can be used for the same problem. As is the case here, for this binary classification, six classification algorithms used commonly for such healthcare problems are explored. All the algorithms are trained, validated, and tested on all the three datasets, window sizes and feature combinations with range of algorithm-specific tuning parameters. Machine learning algorithm and model are two different terms depicting two different things. Model here means a trained algorithm, which is trained over one dataset, a single window of time, and one combination of features and single set of values for algorithm-specific tuning parameters. If any of these ingredients such as dataset, window size, or feature combination is changed, it will be a separate model. If the parameter tuning is ignored at the time of the algorithm training, the total number of models trained and evaluated are:

$$N_{\mathrm{m}} = [A \times S \times W \times (2^F - 1)].$$

The foremost desire in such cases of developing auto-diagnosis solutions for healthcare problems is to build a generic model that can work equally well in all scenarios, but it is very challenging. For example, here, though a single algorithm K-NN could be identified, which performs well for all the body posture scenarios. Even a single window size can be used when the data is collected while sitting or standing. However, it can be seen from Figure 5.4 that the accuracy for the detection of hydration level does not remain the same for K-NN when any of the postures or window sizes is changed. Besides that, from Table 5.6 it can also be observed that even the feature combinations are different in the best-performing models for different scenarios. In such cases, if one model, that is, an algorithm trained on single set of feature combination, tuning parameters, and window size, is used the performance of the model will dramatically change across datasets collected in different body postures and for different interval of times. So a possible solution here can be the identification of a posture and a window size that can be recommended for data collection so the hydration level can be persistently detected with certain accuracy.

In this study, it is found that the best performing K-NN Model can detect hydration level with an accuracy of 87.78% when the data is collected in the sitting posture in samples of 60 seconds. It is also observed that K-NN outperforms other algorithms exceptionally, particularly in sitting and standing posture scenarios for the window size of 60 seconds, and it marginally performs better in a combined data scenario for a window size of 30 seconds. Another interesting aspect observed from Figure 5.4 is, again, that almost all algorithms perform better on sitting data. This confirms the assumptions made earlier that GSR data collected in the sitting posture is more reliable for hydration level detection. The behavior of individual algorithms across window sizes also shows that performance of an algorithm commonly improves with increase in window size in sitting and standing posture, but it decreases in combined data. This trend follows from window sizes of 30 seconds to 60 seconds, but it seems to reverse after that.

In the choice of the selection of an algorithm, performance of the model in terms of accuracy is one criteria. Other important criteria, particularly for developing low energy real-time solutions, are computation and memory cost of the model. If a model involves a complex method of feature extraction or uses an algorithm that needs significant amounts of data in memory for reference, then it can become inefficient for the real-time low energy diagnosis solutions. For example the K-NN keeps training data in memory for the reference, and so it is expensive in terms of memory consumption. So it can be a good solution if the storage memory is not a limitation, for example, if testing is performed off line on a separate computation device instead of on a wearable low energy sensing and diagnosis device. Another possible solution for real-time diagnosis is that the wearable sensing and diagnosis device collect data an send it to server or cloud and data processing so that testing is performed there and results are communicated back to the device. In such a case, the device needs to be connected to cloud or server via the Internet. If there are limitations of memory or in cloud computation capacity, then some other algorithm like SVC may be considered for the data collected for 60 seconds in the sitting posture. In that case, accuracy is only a bit compromised from 87.78% to 81.11%.

5.6 Conclusions

Machine learning can revolutionize healthcare services by providing automated solutions for diagnosis and treatment. In this chapter, with the help of a use case, it is highlighted how machine learning can help in developing auto-diagnosis solutions for hydration level detection, which is important for futuristic healthcare services. It encompasses steps from the data collection of the bio-marker to the development of the machine learning model, commonly important for the development of an auto-diagnosis solution. Maintaining an appropriate hydration level is not only important in sickness but also to promote a healthy lifestyle. Towards that end, it is important to identify bio-markers and develop solutions which can detect the hydration level in an automated manner. Here it has been determined that GSR can be used as bio-marker for hydration level detection in human beings. For this purpose data is collected in two states: hydrated and dehydrated. An appropriate length of time and a suitable body posture are identified in which the data can be collected. A set of features, a machine learning algorithm, and values of the tuning parameters are shortlisted that can help to develop a machine learning model for the detection of hydration level with promising accuracy. It has been found that when the GSR data are collected in sitting posture as samples of 60 seconds then K-NN applied on a specific feature set extracted from that data can detect the hydration level in the human body with an accuracy of 87.78%. A key advantage of the solution proposed here is that it is a non-invasive auto-diagnosis solution for the detection of the hydration level in the human body. The true positive rate in sitting posture for the detection of dehydration is even higher, with a value of 0.95. Hence, if the GSR data are collected for 60 seconds in sitting posture, the K-NN based model can detect with an accuracy of almost 95% if an individual is dehydrated.

References

Aristotelous, P., George Aphamis, Giorgos K Sakkas, Eleni Andreou, Marios Pantzaris, Theodoros Kyprianou, Georgios M Hadjigeorgiou, et al. (2019). Effects of controlled dehydration on sleep quality and quantity: a polysomnographic study in healthy young adults, *Journal of Sleep Research*, 28(3), p.e12662.

Armstrong, L.E. (2005). Hydration assessment techniques, *Nutrition Reviews*, 63(suppl 1), p.S40–S54.

Armstrong, L.E. (2007). Assessing hydration status: the elusive gold standard. *Journal of the American College of Nutrition*, 26(sup 5), p.575S–584S.

Armstrong, L.E. (2005). Hydration assessment techniques, *Nutrition Reviews*, 63(suppl 1), p.S40–S54.

Boucsein, W., D.C. Fowles, S. Grimnes, G. Ben-Shakhar, W.T. Roth, M.E. Dawson, and D.L. Filion. (2012). Publication recommendations for electrodermal measurements. Society for Psychophysiological Research Ad Hoc Committee on Electrodermal Measures, *Psychophysiology*, 49(8), p.1017–1034.

Coniglio, M.A., C. Fioriglio, and P. Laganà. (2020). Water and health. In *Non-Intentionally Added Substances in PET-Bottled Mineral Water*, 1–10. Springer.

Cortés-Vicente, E., D. Guisado-Alonso, R. Delgado-Mederos, P. Camps-Renom, L. Prats-Sanchez, A. Martinez-Domeno, and J. Marti-Fabregas. (2019). Frequency, risk factors and prognosis of dehydration in acute stroke, *Frontiers in Neurology*, 10, p. 305.

Da Silva, H.P., J. Guerreiro, A. Lourenço, A.L.N. Fred, and R. Martins. (2014). Bitalino: A novel hardware framework for physiological computing, International Conference on Physiological Computing Systems)PhyCS), p.246–253.

Davies, K.J.A. (2016). Adaptive homeostasis, *Molecular Aspects of Medicine*, 49, p.1–7.

Evans, G.H., R.J. Maughan, and S.M. Shirreffs. (2017). Effects of an active lifestyle on water balance. In *Nutrition in Lifestyle Medicine*, 281–294. Springer.

Fish, R.M., and L.A. Geddes. (2009). Conduction of electrical current to and through the human body: a review, *Eplasty*, 9.

Fortes, M.B., J.A. Owen, P. Raymond-Barker, C. Bishop, S. Elghenzai, S.J. Oliver, and N.P. Walsh. (2015). Is this elderly patient dehydrated? diagnostic accuracy of hydration assessment using physical signs, urine, and saliva markers, *Journal of the American Medical Directors Association*, 16(3), p.221–228.

Garrett, D.C., N. Rae, J.R. Fletcher, S. Zarnke, S. Thorson, D.B Hogan, and E.C. Fear. (2017). Engineering approaches to assessing hydration status, *IEEE Reviews in Biomedical Engineering*, 11, p.233–248.

Hossin, M., and M.N. Sulaiman. (2015). A review on evaluation metrics for data classification evaluations, *International Journal of Data Mining & Knowledge Management Process*, 5(2), p.1.

Howell, T.A., A. Hadiwidjaja, P.P. Tong, C, Douglass Thomas, and C. Schrall. (2019). Method and apparatus for hydration level of a person, August 15, 2019. US Patent App. 16/270,773.

Khatun, M.F., M.S. Rana, T.A. Fahim, and S.T. Zuhori. (2019). Mathematical models for extracellular uid measurement to detect hydration level based on bioelectrical impedance analysis. In 2019 IEEE Canadian Conference of Electrical and Computer Engineering (CCECE), pages 1–5. IEEE.

Lee, N.E., T.H. Lee, D.H. Seo, and S.Y. Kim. (2015). A smart water bottle for new seniors: internet of things (iot) and health care services, *International Journal of Bio-Science and Bio-Technology*, 7(4), p.305–314.

Manz. F. (2007). Hydration and disease, *Journal of the American College of Nutrition*, 26(sup5), p.535S–541S.

Marra, M., R. Sammarco, A. De Lorenzo, F. Iellamo, M. Siervo, A. Pietrobelli, L.M. Donini, et al. (2019). Assessment of body composition in health and disease using bioelectrical impedance analysis (bia) and dual energy x-ray absorptiometry (dxa): a critical overview, *Contrast Media & Molecular Imaging*, 2019.

Merhej. R. (2019). Dehydration and cognition: an understated relation, *International Journal of Health Governance.*

Pedregosa, F., G. Varoquaux, A. Gramfort, V. Michel, B. Thirion, O. Grisel, M. Blondel, P. Prettenhofer, et al. (2011). Scikit-learn: machine learning in Python, *Journal of Machine Learning Research*, 12, p.2825–2830.

Pross. N. (2017). Effects of dehydration on brain functioning: a life-span perspective, *Annals of Nutrition and Metabolism*, 70(Suppl. 1), p.30–36.

Rizwan, A., N. Abu Ali, A. Zoha, M. Ozturk, A. Alomainy, M.A. Imran, and Q.H. Abbasi. (2020). Non-invasive hydration level estimation in human body using galvanic skin response, *IEEE Sensors Journal*, 20(9), p.4891–4900.

Rizwan, A., A. Zoha, R. Zhang, W. Ahmad, K. Arshad, N. Abu Ali, A. Akram Alomainy, et al. (2018). A review on the role of nano-communication in future healthcare systems: a big data analytics perspective, *IEEE Access*, 6, p.41903–41920.

Sagner, M., A. McNeil, P. Puska, C. Auffray, N.D. Price, L. Hood, C.arl J Lavie, et al. (2017). The p4 health spectrum: a predictive, preventive, personalized and participatory continuum for promoting healthspan, *Progress in Cardiovascular Diseases*, 59(5), p.506–521.

Schaffer, C. (1993). Selecting a classification method by cross-validation, *Machine Learning*, 13(1), p.135–143.

6

Information Retrieval from Electronic Health Records

Meshal Al-Qahtani, Stamos Katsigiannis and Naeem Ramzan

School of Computing, Engineering and Physical Sciences, University of the West of Scotland

Advances in computing encouraged the adoption of computer systems in numerous applications. In the health domain, the adoption of computer systems enables the introduction of better services, the provision of reliable services, and the reduction of human errors. Generally, data in computer systems are stored in coded format. However, in health databases some data cannot be coded, such as doctors, comments; hence, they are stored in the form of free text. Available literature has demonstrated that such free text contains invaluable information. However, extracting information from the free text portion of health databases is a challenging task due to the complexity of the stored data. Latent semantic indexing (LSI) is an information retrieval (IR) technique that has proven its effectiveness in extracting information from health databases, as it is able to identify the semantics of the terms within and across the documents within the database. However, LSI has a major limitation, which is its inefficiency when extracting information from large-scale document collections. In this chapter, two enhancements of the LSI method are proposed and evaluated in order to overcome this limitation. The proposed distributed LSI and parallel LSI methods were applied to an artificial electronic health records database (EMRbots) and were evaluated in terms of time complexity, recall, and precision.

6.1 Introduction

In the health domain, IR is performed for various reasons, such as expanding understanding about various medical issues with a specific end goal to improve the provided health services (Meystre et al. 2008; Shah, Martinez, and Hemingway 2012; Wang et al., 2012). There is sophisticated inter-relation between electronic health records (EHRs) in the health databases. For example, different free-text records may refer to different medical episodes, which may be included in the family history. The increased number of EHRs in the database increases the complexity of the IR process (Fu et al. 2014). Besides that, there are numerous components that are of concern while considering the health-related points-of-interest of any patient, such as age, gender, and family history. Any IR methodology ought to scale such that it joins these components, while taking into consideration the computational cost, as well as the desired levels of precision and efficiency.

Engineering and Technology for Healthcare, First Edition.
Edited by Muhammad Ali Imran, Rami Ghannam and Qammer H. Abbasi.

Simple search and non-semantic approaches are not suitable for IR in medical text, as contextual information can affect the results' accuracy (Khan. Doucette, and Cohen; 2011 Rahimi and Vimarlund 2007). IR approaches intended for EHRs ought to be executed in view of natural language processing (NLP) to guarantee that they provide the desired level of reliability and accuracy (Ding 1999). NLP-based methodologies address the semantics of the words inside the text, and the outcomes are decided in view of the real semantics of the text rather than on individual keywords (Moen et al. 2015). LSI demonstrated its suitability for IR in the health domain as it addresses the semantics of the terms inside the text. However, LSI is inefficient at processing continuously scaling databases (Deerwester et al. 1990). For example, in The Health Improvement Network (THIN) database, there are more than 209,000 free-text fragments for 4000 patients, with about 35,000 recognized terms (THIN). The target of this chapter is to improve the capacity of the LSI method to process databases of substantial scale and accomplish a specific level of the desired accuracy.

Information extraction from such data requires a considerable amount of time and memory. One of the best solutions for this issue is isolating the expansive dataset into smaller subsets that can be conveyed into numerous clusters and taken care of effortlessly. Isolating the subsets with no principles is not efficient, since this would eliminate concealed relations between elements of the dataset. To address this issue, the best approach for partitioning the dataset and keeping the concealed relations between their elements is clustering. Clustering is a method for separating data into groups of related elements, such that elements in each group have similar properties.

6.2 Methodology

LSI matches a user query to a term document matrix (TDM) and identifies those documents that best match the query (Guillet and Hamilton, 2007). The TDM holds the number of occurrences of the terms in each document in the documents collection. Basically, LSI performs the following steps:

- **Pre-processing:** Each document in the document collection is processed, such that only those terms that contribute to the information retrieval process appear in the TDM. At this stage, stop words like "a," "the," and "in" are excluded. In general, the time required to perform this step is proportional to the number of documents in the document collection and the size of these documents.
- **Processing:** When a user query is provided, it is possible that the terms in the query may have different weights or the same weight depending on the importance of the term. The TDM is refined using singular value decomposition (SVD). The refining is used to increase the relevancy probability such that documents that have the same semantics are grouped together. Then the relevancy of each document in the TDM to the query is identified using the cosine similarity.
- **Post-processing:** A threshold is used to set the minimum relevancy for a document to appear in the results, and the documents that meet that threshold are listed as results.

The pre-processing stage is expected to have high memory and computational requirements that scale as the documents and terms increase, something that significantly affects

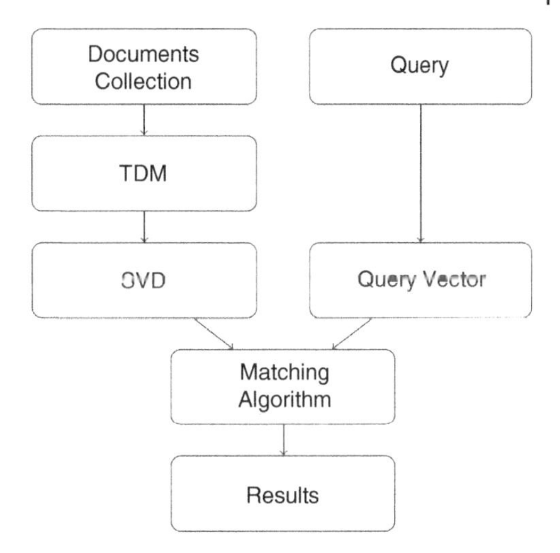

Figure 6.1 Basic LSI Technique.

the efficiency of the IR process. Different methods have been proposed to enhance the efficiency. For example:

- The use of planned stop word lists relatively enhances the efficiency of the IR process.
- LSI is able to address the synonym and polysemy of the terms, which enhances the accuracy of the IR process (Jaber, Amira, and Milligan 2012).

The IR process of LSI is outlined in Figure 6.1. However, LSI's efficiency is reduced when performing IR from large document collections. To address the issue of efficiency in large-scale databases, two enhancements of the LSI methods are proposed: the parallel LSI and the distributed LSI.

6.2.1 Parallel LSI (PLSI)

Straightforward search and non-semantic IR procedures are not appropriate for the health domain, as contextual information can affect accuracy (Fu et al. 2014). Furthermore, the representation of the patients and of the medical notes within the database also affects efficiency.

For IR purposes, it is important to identify a certain methodology for the representation of the patient in the TDM, such that meaningful information is extracted. Thus, if the patient is considered as a document, all the free-text records for the patient are collected as a single document. However, those free-text records usually refer to different medical events, and these events might be unrelated, a problen that introduces noise into the IR process. On the other hand, if each free-text record is considered as a document, it is necessary to build a large-size TDM. For example, a lab test or a symptom that is shared between different medical cases might provide false information about the relevancy of such documents for the user's information needs.

Accordingly, those free-text records that fall within the same medical status are relatively closer to each other. Based on that, multiple TDMs can be built, each one of them related to a certain medical status. This is expected to minimize the noise of unrelated medical events and increase the accuracy of the LSI.

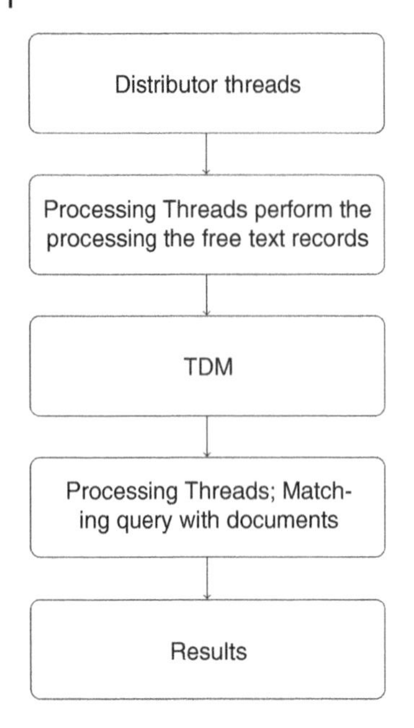

Figure 6.2 Parallel LSI.

An additional challenge to the IR process within the health domain is the possibility that certain records may have a family history discussion; for example, the doctor might address the previous history of a family member for an expected medical case. The pre-processing stage should be able to discover such cases and ensure that these terms do not appear in a patient's document in the TDM.

Based on that, a parallel version of the LSI IR technique has been proposed as shown in Figure 6.2. The proposed algorithm parallelizes the pre-processing of the free-text records of the database, by distributing the free-text records on multiple processing threads that are running concurrently. The calculation of the relevancy between the query and the documents in the document collection is also parallelized to further boost the efficiency of the IR process.

The pre-processing of the free-text records considers synonymy and polysemy, using proper alternatives of the same term to enhance the accuracy of LSI. The parallel processing of the synonymy and polysemy significantly reduces the time requirements of the IR process. The pre-processing is designed to utilise a general English dictionary to handle spelling errors and abbreviations, which are added to the TDM if they contain no spelling errors. Terms that contain spelling errors are not added to the TDM.

After the pre-processing of all the free-text records, each document in the TDM represents the set of free-text records that refer to the same patient. Then, SVD is applied to the TDM. The three basic matrices U, S, and V are identified and the rank of the TDM is identified based on the number of the non-zero values in S. A set of processing threads calculates the relevancy between the query and each document in the TDM. Currently, all the terms in the query are given the same weight. However, it is possible to use multiple weights according to the IR process requirements.

6.2.2 Distributed LSI (DLSI)

Clustering is meant to discover the structure of data (Zhang, Yoshida, and Tang 2011). Data clustering, defined as a statistical classification technique, can be used to find if objects of a population can be grouped into subsets that share common characteristics.

There are three main methods for clustering data: compression, natural classification, and underlying structure (Xu and Wunsch 2005). The main categories of clustering are hierarchical methods, overlapping clustering, partitioning algorithms, ordination techniques, and so on (Ball and Hall 1967). The proposed distributed LSI uses the K-means algorithm for data clustering.

K-means clustering is one of the most popular clustering algorithms and various parallel procedures have been proposed to accelerate the K-means algorithm (Stoffel and Belkoniene 1999). Its advantages include:

- Simple implementation
- Guaranteed convergence
- Support for clusters of various shapes and sizes

However, the K-means algorithm has some drawbacks:

- There is no general effective strategy for deciding the underlying number K of groups. A general solution for this is to arbitrarily run the algorithm a few times for different K. Another solution is by utilizing a system, called ISODATA, introduced in (Xu and Wunsch 2005), which can progressively modify the amount of clusters by consolidating and partitioning clusters, as indicated by some predefined threshold.
- Results are sensitive to initialization. Various methods for selecting the initial values of the centroids have been proposed (Celebi, Kingravi, and Vela 2013).
- K-means is sensitive to outliers. If there is an element that is too far from cluster centroids, it's compelled to join the closest cluster.

In the context of LSI, data elements can represent the SVD of each document in the documents collection. Clustering the SVD of the documents accelerates the IR process, as the relevancy of the user's query can be measured to the cluster that it falls within, which significantly minimizes the time requirements of the IR process.

From another perspective, the clustering process has a major advantage in the IR process, which is the removal of noise. As all those documents that are of high relevance to each other are located in the same cluster, the IR process will not consider any document that falls in another cluster that did not meet the user's query requirements.

The proposed DLSI algorithm attempts to provide appropriate data distribution among clusters to improve the managing of large-scale datasets. To distribute data among clusters, the documents are distributed in a way that related documents are grouped together, using a Balanced K-means clustering, as follows:

1. The vector $\vec{t} = [t_1, t_2, ..., t_n]$ of the distinct user query terms t_j and the vectors $\vec{d_k} = [f_{1k}, f_{2k}, ..., f_{nk}]$, f_{jk} being the tf-idf of t_j for document k, are created.
2. K documents are selected, with each document representing the initial group centroid.

3. Each document is assigned to the nearest group centroid, with respect to balancing the number of documents in each centroid, or balancing the number of terms in each centroid. Equations 6.1 and 6.2 are used for assigning the documents to the closest centroid.

$$\text{Cosine distance} = \frac{\vec{A}\vec{B}}{||\vec{A}||\ ||\vec{B}||} = \frac{\sum_{i-1}^{n}(A_iB_i)}{\sqrt{\sum_{i=1}^{n}(A_i)^2}\sqrt{\sum_{i=1}^{n}(B_i)^2}} \tag{6.1}$$

$$\text{Similarity(doc, centroid)} = \text{Cosine distance} \cdot P_{term} \cdot P_{doc} \tag{6.2}$$

4. The positions of K centroids are recalculated after all the documents have been assigned.
5. Steps 2 and 3 are repeated until the centroids $C = \{c_1, c_2, ..., c_k\}$, with c_i being the ith cluster, no longer change.
   This produces a separation of the documents into groups. At this stage, the documents are distributed into different clusters. TDM and SVD are calculated for the documents in each cluster, with one SVD for each cluster. The SVD calculation process is the same as for the SVD in the LSI algorithm.

The user query vector $q = [f_1, f_2, ..., f_n], f_j$ being the average term matrix frequency of query term t_j, is then calculated and the cosine distance (equation 6.1) between the query vector and each centroid c_i is computed as the similarity F_i:

$$F_i = \text{Similarity}(c_i, q) \tag{6.3}$$

Then the new ranks for each document are computed as:

$$R_{(i,d)=r_{(i,d)} \cdot F_i} \tag{6.4}$$

where R is the new rank for document d in the cluster i, and r is the old rank for document d in the cluster i.

Then, documents are sorted according to the new rank value and the results are retrieved accordingly. An overview of the proposed DLSI technique is shown in Figure 6.3.

6.3 Results and Analysis

The parallel and distributed proposals of LSI were evaluated in terms of time requirements and accuracy against the traditional LSI. The evaluation was performed using an artificial database of patient records, EMRbots (Bahrami and Singhal 2015). EMRbots is a database that is used for evaluation purposes and has the same structure as real electronic health records (EHRs). It has been developed by the BOTS organization for testing scientific research (Bahrami and Singhal 2015).

In this chapter, two data samples of EMRbots have been used; one contained 2500 free-text records for comparison purposes, and the other one contained 100,000 free-text records for testing the proposed DLSI.

To evaluate the balanced K-means clustering, the 2500 documents from EMRbots were distributed among five clusters. Figure 6.4 shows that each cluster had almost the same number of documents. Furthermore, the balanced K-means not only distributed the documents almost equally across the clusters, it also balanced the number of terms. However, it

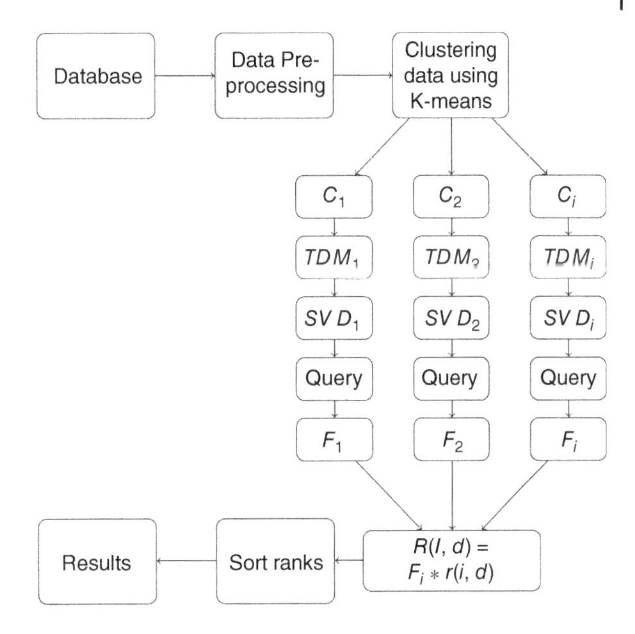

Figure 6.3 Distributed LSI.

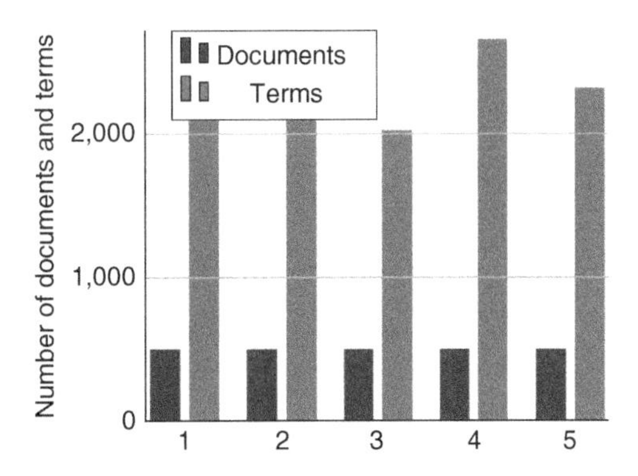

Figure 6.4 Distribution of 2500 documents in Five Clusters Using the Balanced K-means Clustering.

is too difficult to have the same number of terms per cluster due to the variation of terms per document.

Figure 6.5 shows the comparison results for LSI, multi-threaded LSI (PLSI) and DLSI, in terms of time requirements of pre-processing and query search for the 2500 documents. The results show that DLSI requires the least time for pre-processing and query search. The results in Figure 6.6 indicate that DLSI demonstrated a significant reduction in SVD processing time compared to LSI.

The results show that the DLSI enhances the performance of LSI in terms of time requirements, enabling the use of DLSI in large-scale databases. To evaluate this argument, DLSI was evaluated on 100,000 free-text records from the EMRbots database. Figure 6.7 shows the time required for pre-processing, post-processing, and query search for 100,000 records using DLSI. The documents were distributed among 100 clusters in order to overcome the memory-limitation issue. The results indicate that DLSI is suitable for IR in large databases.

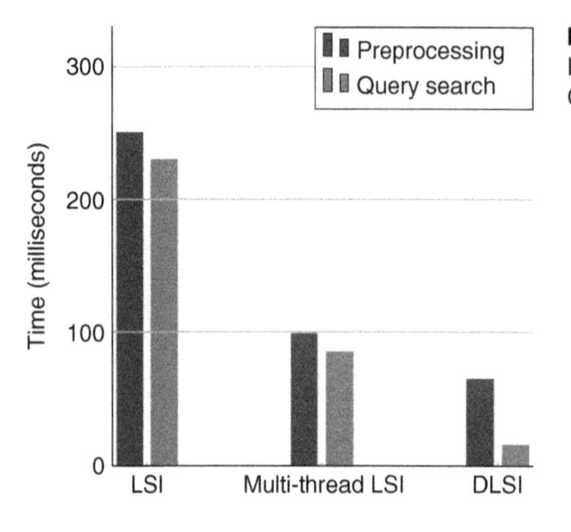

Figure 6.5 Comparison between LSI, Multi-Thread LSI and DLSI in Terms of Computational Time.

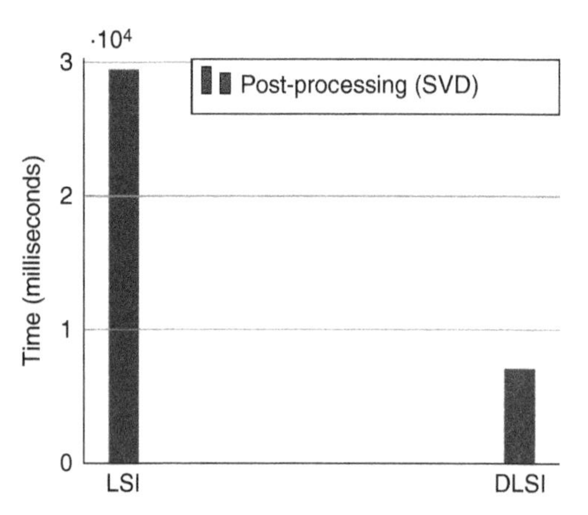

Figure 6.6 Post-Processing Comparison between LSI and DLSI.

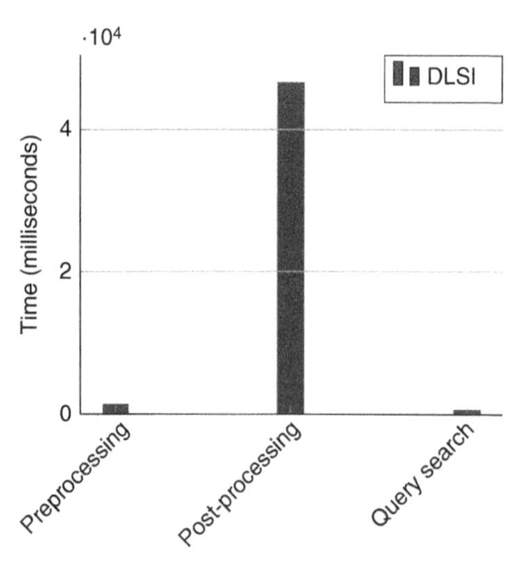

Figure 6.7 Performance Evaluation for DLSI on a Large-Scale Database (100,000 Documents).

Another challenge is finding a suitable K value for the K-means clustering in order to speed up the retrieval while maintaining a minimum level of accuracy. As the value of K decreases, the retrieval time decreases, but the accuracy decreases as well.

The best value of K can be found using binary search. Initially the value of K is the maximum value. Every time the result for a lower K value is compared to the current value, the accuracy is calculated by equation 6.5. If the accuracy for the lower K value matches the current value, the current value is set to be the lower value. This process is iterated until the result for a new value of K does not match the results of the current value, which is then considered the best value for K.

$$Accuracy = \frac{\text{Similarity for best document of query K}}{\text{Similarity when K has maximum value}} \tag{6.5}$$

The results indicate that a 90% accuracy score was obtained when K=31, while a perfect accuracy score of 100% was obtained when the value of K was over 900. Figure 6.8 shows the accuracy for multiple K values. In addition, the least processing time is obtained for a K value of 31, as shown in Figure 6.9.

The efficiency of DLSI was compared to the parallel LSI. The results show that DLSI achieved better precision and recall compared to parallel LSI, as shown in Figure 6.10 for

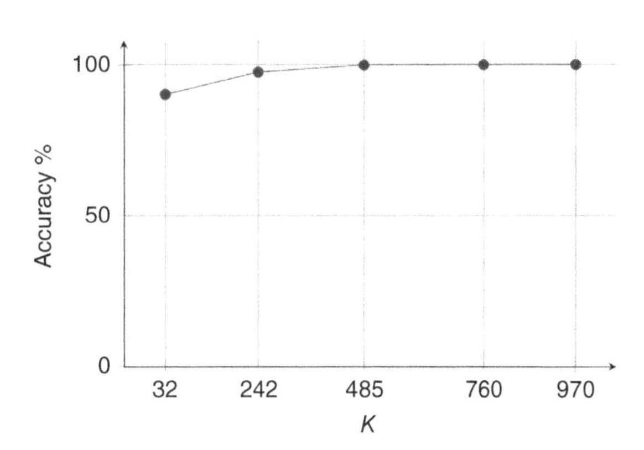

Figure 6.8 Accuracy vs. K Value.

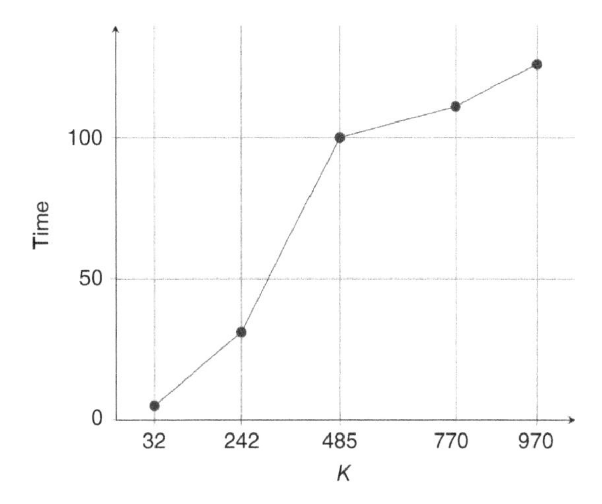

Figure 6.9 Time vs. K Value.

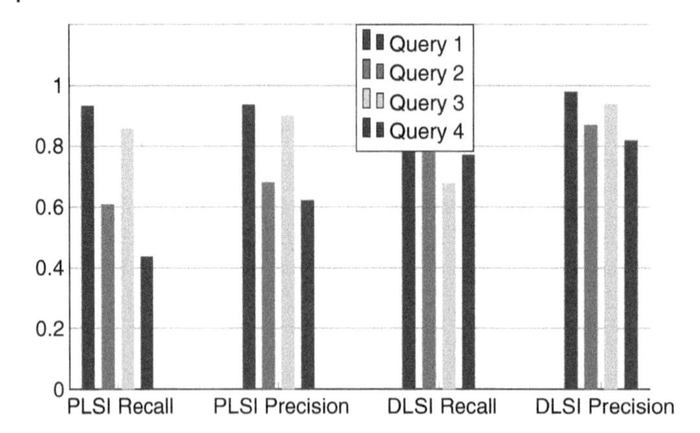

Figure 6.10 Comparison Results (Recall and Precision) between LSI and DLSI.

four queries applied to the sample of 2500 documents from the EMRbots database. The results indicate that DLSI significantly enhances the efficiency of the LSI.

6.4 Conclusion

This chapter has demonstrated the efficiency of two enhanced versions of latent semantic indexing (LSI) for information retrieval (IR) from free text in large-scale datasets. The experimental evaluation on an artificial electronic health records database showed that both the proposed parallel LSI and distributed LSI outperformed the standard LSI, requiring significantly lower computational times, thus allowing the use of LSI on large-scale datasets.

References

Bahrami, M., and M. Singhal. (2015). A dynamic cloud computing platform for ehealth systems, in *2015 17th International Conference on E-health Networking, Application Services (HealthCom)*, p. 435–438, Oct.

Ball, G.H., and D.J. Hall. (1967). A clustering technique for summarizing multivariate data, *Behavioral Science*, 12(2), p. 153–155.

Celebi, M.E., H.A. Kingravi, and P.A. Vela. (2013). A comparative study of efficient initialization methods for the k-means clustering algorithm, *Expert Systems with Applications*, 40(1):200–210, 2013.

Deerwester, S., S.T. Dumais, G.W. Furnas, T.K. Landauer, and R. Harshman. (1990). Indexing by latent semantic analysis, *Journal of the American Society for Information Science*, 41(6), p. 391–407.

Ding, C.H.Q. (1999). A similarity-based probability model for latent semantic indexing, in Proceedings of the 22nd Annual International Conference on Research and Development in Information Retrieval *(SIGIR '99)*.

Fu, Z., X. Sun, N. Linge, and L. Zhou. (2014). Achieving effective cloud search services: multi-keyword ranked search over encrypted cloud data supporting synonym query, *IEEE Transactions on Consumer Electronics*, 60(1), p. 164–172.

Guillet, F., and H.J. Hamilton. (2007). *Quality Measures in data mining*, Berlin; Heidelberg: Springer-Verlag.

Jaber, T., A. Amira, and P. Milligan. (2012). Enhanced approach for latent semantic indexing using wavelet transform, *IET Image Processing*, 6(9), p. 1236–1245.

Khan, A., J.A. Doucette, and R. Cohen. (2011). An ontological approach for querying distributed heterogeneous information systems 76–88. In *CSWS*, 2011.

Meystre, S.M., G.K. Savova, K. Kipper-Schuler, and J.F. Hurdle. (2008). Extracting information from textual documents in the electronic health record: a review of recent research. *Yearbook of medical informatics*, p. 128–44.

Moen, H., MF. Ginter, E. Marsi, L.-M. Peltonen, T. Salakoski, and S. Salanter. (2015). Care episode retrieval: distributional semantic models for information retrieval in the clinical domain, *BMC Medical Informatics and Decision Making*, 15(S2).

Rahimi, B. and V. Vimarlund. (2007). Methods to evaluate health information systems in healthcare settings: a literature review, *Journal of Medical Systems*, 31, p. 397–432.

Shah, A.D., C. Martinez, and H. Hemingway. (2012). The freetext matching algorithm: a computer program to extract diagnoses and causes of death from unstructured text in electronic health records, *BMC Medical Informatics and Decision Making*, 12(88).

Stoffel, K., and A. Belkoniene. (1999). *Parallel k/h-means clustering for large data sets, in Proceedings of the 5th International Euro-Par Conference on Parallel Processing, Euro-Par '99*. London: Springer-Verlag.

THIN. The health improvement network. https://www.the-health-improvement-network.co.uk/. Accessed: 2019-12-02.

Xu, R., and D. Wunsch. (2005). Survey of clustering algorithms, *IEEE Transactions on Neural Networks*, 16(3), p. 645–678.

Wang, Z., A.D. Shah, A.R. Tate, S. Denaxas, J. Shawe-Taylor, and H. Hemingway. (2012). Extracting diagnoses and investigation results from unstructured text in electronic health records by semi-supervised machine learning, *PLOS ONE*, 7(1), p. 1–9.

Zhang, W., T. Yoshida, and X. Tang. (2011). A comparative study of TF*IDF, LSI and multi-words for text classification, *Expert Systems with Applications*, 38(3), p. 2758–2765.

7

Energy Harvesting for Wearable and Portable Devices

Rami Ghannam[1], You Hao[2], Yuchi Liu[3] and Yidi Xiao[4]

[1]*Electronics and Nanoscale Engineering, University of Glasgow*
[2]*Electronic and Electrical Engineering, University of Glasgow*
[3]*Electronics and Nanoscale Engineering, University of Glasgow*
[4]*Electronic and Electrical Engineering, University of Glasgow*

The aim of this chapter to provide an overview of the different techniques that can be used to harvest energy for portable and wearable electronic devices. These technologies rely on converting ambient energy, or energy from the body into useful electricity. We discuss each technology's working principles, as well as its merits for satisfying the power demands of wearable electronic devices.

7.1 Introduction

Thanks to advancements in packaging and nanofabrication, it is now possible to embed various microelectronic components into a small area, on a flexible substrate and at a relatively low cost. This trend in miniaturization has enabled wearable electronic devices to be integrated with sensors, actuators, and wireless communications technology. Consequently, wearable electronic devices and health trackers are currently in high demand. Moreover, they can improve the quality of life as well as the life expectancy of many chronically ill patients, provided that certain biological signs are accurately monitored. Moreover, they can be used to enable communication with other electronic devices, either via gesture recognition (Liang, Ghannam, and Heidari 2018; Liang et al. 2019) or via eye movement (Yuan et al. 2020). However, to ensure that these devices can truly improve a patient's quality of life, novel energy harvesting solutions that are non-obstructive to their wearer are required. First, it is important to understand how much power is typically required to support the various components in a wearable device. A summary of these power requirements is shown in Table 7.1 (Vullers et al. 2009b).

From the data in Table 7.1, it is clear that low amounts of power (less than 50 mW) are required to meet the needs of these wearable devices. Often, flexible chemical-cell batteries have been used to partially fulfil these needs. However, these batteries need regular charging and eventual replacement. Additionally, the limited available energy restricts the size and weight of the wearable device. Consequently, the wearer will need to take off the

Engineering and Technology for Healthcare, First Edition.
Edited by Muhammad Ali Imran, Rami Ghannam and Qammer H. Abbasi.

Table 7.1 Power Consumption of Electronic Devices Vullers et al. (2009b)

Device type	Power consumption (W)
Smart phone	1 W
MP3 Player	50 mW
Hearing aid	1 mW
Wireless sensor node	100 μW
Cardiac pacemaker	50 μW
watch	5 μW

device for charging purposes, which makes it ineffective for continuous health monitoring. Eliminating this energy storage reservoir in a wearable device is often impossible, since peak currents and continuous device operation are needed (Vullers et al. 2009b). Thus, combining an energy harvester with a compact-sized rechargeable battery (or super capacitor) appears to be the best option for achieving energy autonomy. In this case, electrical energy generated from the surroundings or from the human body can open up new possibilities for powering body-worn devices. In this chapter, we will focus on the electromagnetic, piezoelectric, and thermal power harvesting techniques to make wearable electronic devices self-sustainable.

7.2 Energy Harvesting Techniques

7.2.1 Photovoltaics

Harvesting energy from light has been used for powering portable consumer products that include calculators, wristwatches, and toys. Here, photovoltaic (PV) cells are used to convert light or the sun's energy into useful electricity. Ultimately, semiconductor materials are commonly used for the purpose of producing currents and voltages as a result of the absorption of sunlight, which is a phenomenon known as the photovoltaic effect. Most solar cells are fabricated from either monocrystalline or polycrystalline silicon (Si) materials. In its most basic form, a solar cell consists of a *pn* junction diode. Typical solar cell efficiencies range from 18% for polycrystalline to 24% from highly efficient monocrystalline technologies. These high end PV devices typically include special light-trapping structures that absorb as many of the incident photons as possible Abdellatif et al. (2015; Michel, Moll, and Ghannam 2018).

Figure 7.1 shows the basic operating principles of a solar cell. Both drift and diffusion of carriers takes place across the depletion region of width, W. The built-in electric field E_0 in this depletion layer prevents further diffusion of minority carriers. The finger electrodes on the surface of the n-type semiconductor material allows light to penetrate into the device. Furthermore, these electrodes result in a small series resistance. The photogenerated electron-hole pairs in the depletion region become separated by the built-in electric field, E_0. Through the process of drift, electrons reach the neutral n-region and make it negative by an amount of charge $-q$. Similarly, holes drift to the p-region, which effectively turns that region more positive. Consequently, an open circuit voltage, V, develops between

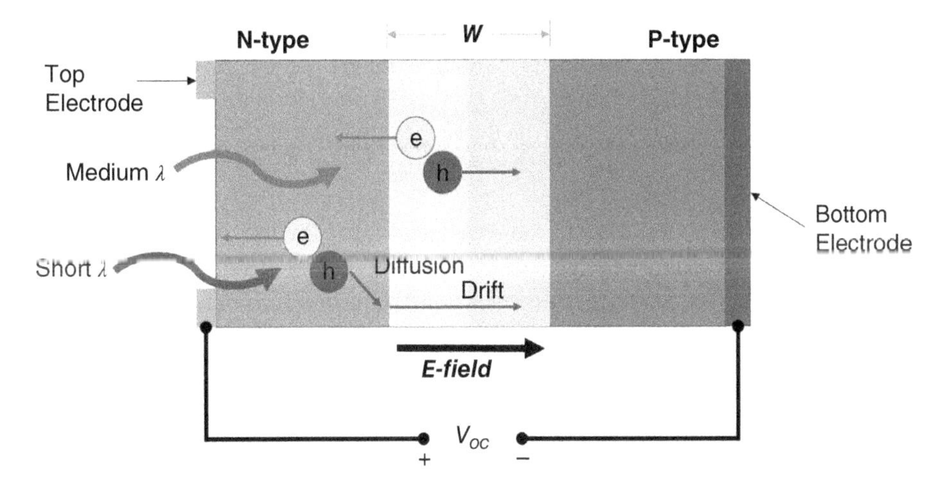

Figure 7.1 Schematic diagram of PV cell. Reproduced with permission from Ghannam, Klaine, and Imran (2019).

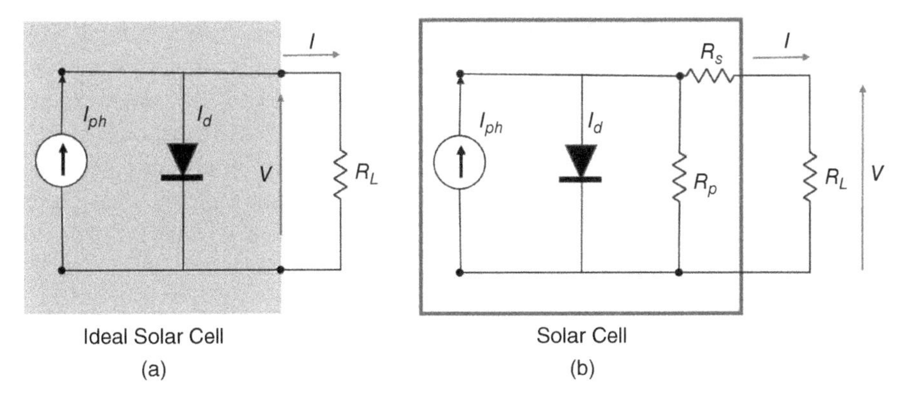

Figure 7.2 Equivalent circuit representation of an (a) Ideal PV cell, and (b) Practical PV cell. Reproduced with permission from Ghannam, Klaine, and Imran (2019).

the terminals of the device, whereby the p-region is positive with respect to the n-region (Ghannam, Klaine, and Imran 2019; Kasap 2006).

In practice, photogenerated electrons need to travel across a semiconductor region in order to be collected by the nearest electrode. Consequently, an effective series resistance, R_s, is introduced in the photovoltaic circuit. Similarly, photogenerated carriers flow through the crystal surfaces or through grain boundaries in polycrystalline devices. These effects can be described in terms of a shunt resistance, R_p, which drives photocurrent away from the load, R_L. Consequently, the equivalent electrical circuit representation of a typical solar cell can be modelled as shown in Figure 7.2.

PV cells are usually classified as crystalline or non-crystalline. There are also four important parameters that define the performance of a solar cell. These are the open circuit voltage V_{OC}, short circuit current I_{SC}, maximum power output P_{mp} and the efficiency η. Thus, from Figure 7.2, an expression for the total output current of the cell, I, can be deduced:

$$I = I_{ph} - I_0(e^{q(V+IR_s)/\eta_o kT} - 1) - ((V + IR_s)/R_p) \tag{7.1}$$

where q is the electric charge, V is the voltage, k is the Boltzmann constant, and T is the cell temperature in Kelvins, K. Consequently, we can determine the $I - V$ characteristics of an SC as a function of input solar radiation, series resistance, and shunt resistance. Moreover, I_o is the reverse saturation current, and the diode ideality factor, η_o, typically depends on the type of solar cell technology used. For the case of monocrystalline silicon, this is usually $\eta_o = 1.2$.

The efficiency of a solar cell indicates how well it converts the sun's energy into useful electricity. Typical efficiencies of crystalline silicon solar cells are in the range of 18% to 25% (Green 2019). Multi-junction PV cells consist of multiple single-junction PV cells that are stacked in tandem with one another. These cells consist of different semiconducting materials and are now capable of achieving 38% efficiency (Green et al. 2006). Thus, the power density that can be harvested is in the range of $18 - 25mW/cm^2$ for outdoor applications. Since the intensity of light is lower indoors, the energy density for these cells is typically $< 10\mu\ W/cm^2$ (Adu-Manu et al. 2018). Despite their high efficiency, crystalline silicon solar cells are not ideal for wearable applications due to their rigid nature. However, there is currently interest in using these cells for implantable electronic applications (Zhao et al. 2018, 2019).

On the other hand, there is rapid development in flexible non-crystalline solar cells, which enable them to be used in wearable electronic applications. Examples include dye sensitised, quantum dot, perovskite, organic, Cadmium Telluride (CdTe), Copper Indium Gallium Selenide (CIGS), and amorphous silicon (a-Si) solar cells. The cell efficiency of these technologies is summarized in Table 7.2. The advantages and disadvantages of these technologies are also mentioned. Currently, flexible multijunction solar cells have been developed for portable electronic applications, which can achieve efficiencies of 30.4%

Table 7.2 Comparison of different PV Technologies (Polman et al. 2016)

Technology	Cell efficiency	Strengths and weaknesses
Multijunction Solar Cell	42.1	High power density, long lifespan, but expensive for wearable applications.
Monocrystalline Si	25.6	Abundant, long lifespan, but not flexible.
Polycrystalline Si	21.6	Abundant, low-cost, not flexible material.
CIGS	21.7	Comparable power density to crystalline technology, flexible, but expensive.
CdTe	21.5	Low cost, flexible, but toxic due to Cadmium presence.
Perovskite	21.0	Low cost, flexible, high power density, but unstable and poor lifespan.
CZTS	12.6	Flexible, but low power density.
Dye Sensitised Solar Cell	11.9	Low cost, flexible, but poor stability.
Amorphous Si	11.4	Low degradation with temperature, low cost, flexible, but low power density.
Organic	11.5	Flexible, semi-transparent, but poor stability and lifespan.
Quantum Dots	9.9	Low cost, flexible, but poor power density.

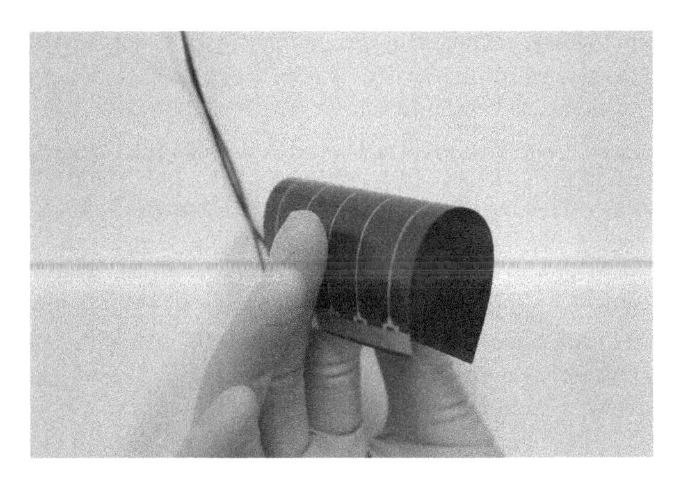

Figure 7.3 Flexible PV cell for wearable applications.

(Cardwell et al. 2017; Chan et al. 2018; Kirk et al. 2018; Scheiman et al. 2014; Stender et al. 2015; Trautz et al. 2013).

The Soliband is an example of a wearable wristband device that makes use of flexible solar cells (Dieffenderfer et al. 2014), such as the cell shown in Figure 7.3. A battery was used to provide a power of 13.7 mW for a duration of 4 hours. Other examples of solar powered wristband devices have been developed by ETH Zurich's Integrated Systems Laboratory, which is led by Professor Luca Benini (Kartsch et al.; Magno et al. 2016a,b, 2019; Polonelli et al. 2018).

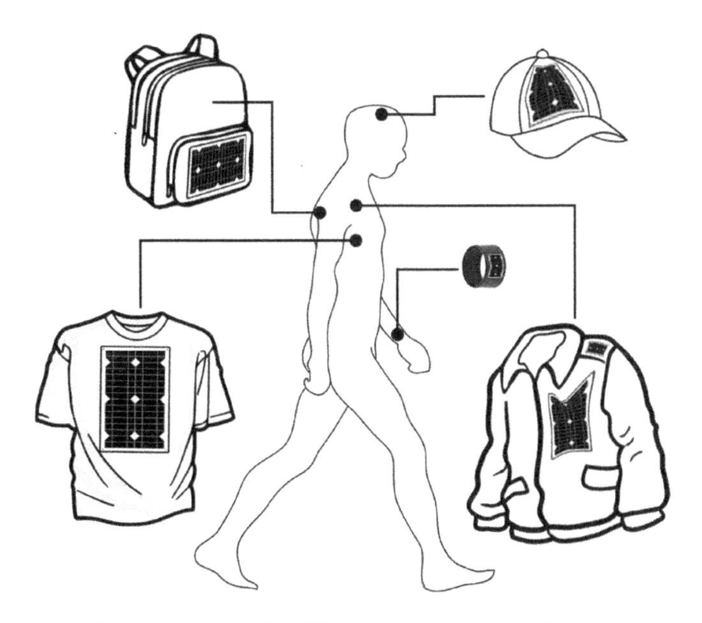

Figure 7.4 PV cells used in different wearable applications. These applications may include bags, T-shirts, jackets, wristbands and caps.

Furthermore, there is currently plenty of interest in integrating polymer solar cells in woven fabrics (Cho et al. 2019; Jeong et al. 2019; Liu et al. 2018b; Zhang et al. 2014), and particularly with dye sensitized solar cells woven in fabrics (Liu et al. 2018a, 2019a,b; Mueller et al. 198; Pu et al. 2016; Song et al. 2019; Yun et al.). Thus, the free and abundant nature of solar electricity makes it ideal for powering electronic devices. Recent progress in advancing device efficiencies has enabled large power densities to be achieved. However, solar energy is unpredictable, and is location as well as time dependent. Consequently, solar electric devices need to be combined with power conditioning circuitry to ensure their appropriate use in a wearable system (Htet et al. 2018). An overview of the energy storage technologies suitable for integration into flexible power systems is provided in Ostfeld and Arias (2017). We will therefore next introduce piezoelectric energy devices and how they can be used to harvest energy from a wearer's movement.

7.2.2 Piezoelectric Energy Harvesting

In this section, we summarize some piezoelectric power harvesting techniques to convert mechanical daily motions or organ vibrations into electrical energy, which can be used to power wearable electronics. Novel piezoelectric materials and devices have been developed to ensure comfortability, portability, and high power output.

Equation (7.2) shows the basic principle of piezoelectric effect.

$$\begin{aligned} S &= s^E T + d^t E \\ D &= dT + \varepsilon^T E \end{aligned} \tag{7.2}$$

Where S is the strain tensor, T is the stress tensor, E is the electric field, D is the electric displacement, s^E is the compliance under a zero or constant electric field, ε^T is the dielectric permittivity under a zero or constant stress, d and d^t are the direct and reverse piezoelectric coefficients Hehn and Manoli (2015). Figure 7.5 shows the piezoelectric effect in both the inverse and direct methods. In this case, the direct piezoelectric method converts mechanical energy to electrical energy, which is well-suited for power harvesting. Examples of piezoelectric materials that are widely used for energy harvesting applications include Zinc Oxide (ZnO), Barium Titanate (BaTiO3), Polyvinylidene Fluoride (PVDF) and Zirconate Titanate (PZT).

The different methods for harvesting energy from the body are shown in Figure 7.6. Until now, harvesting kinetic energy from walking is still the most promising piezoelectric technique, due to the large amounts of power that can be generated. According to Starner's calculation, a man weighing 68 kg will generate around 67 W when walking at a speed of two steps per second (Starner 1996).

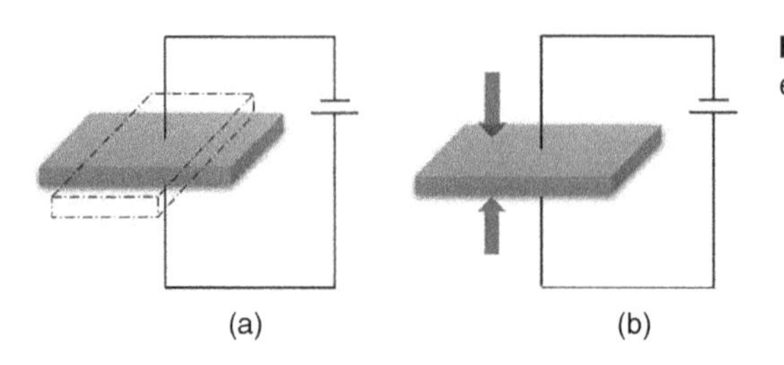

Figure 7.5 (a) Inverse piezoelectric effect, (b) Direct piezoelectric effect.

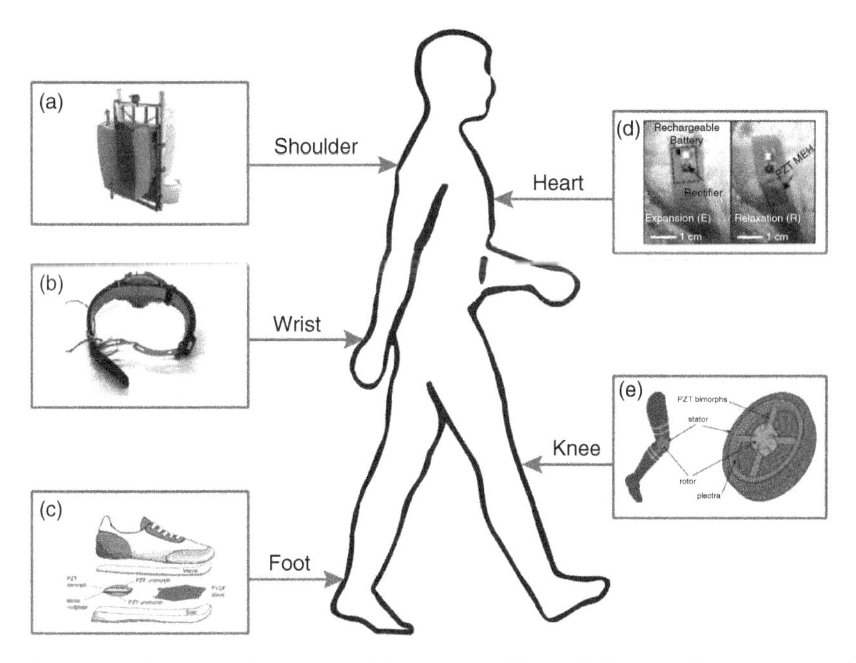

Figure 7.6 Piezoelectric energy could be harvested from(a) Shoulder (Granstrom et al. 2007), (b) Wrist (Jung et al. 2015), (c) Foot (Shenck and Paradiso 2001), (d) Heart (Dagdeviren et al. 2014), (e) Knee (Pozzi and Zhu 2011).

The first demonstration of harvesting piezoelectric energy from human walking was achieved in 1996 by Shenck and Paradiso from the MIT Media Laboratory (Shenck and Paradiso 2001). They designed two different insoles based on the PVDF and PZT piezoelectric materials. One of their designs was based on a flexible PVDF bimorph stave under the insole to capture the dissipated energy from the bending ball of the foot. The average amount of generated power was 1.3 mW for a walking pace of 0.9 Hz and a 250 $k\Omega$ load. Their second method relied on mounting a semiflexible PZT dimorph, which consisted of two back-to-back, single-sided unimorphs under the heel. The output power for same walking pattern was 8.4 mW in a 500 $k\Omega$ load.

Moreover, Jingjing Zhao suggested improvements to the structure of the PVDF piezoelectric insole to reduce wearer discomfort in 2014. The average amount of generated power was 1 mW for a walking pace of 1 Hz. This power was subsequently increased to 50 mW by integrating the power harvesting unit with a direct current power supply. This setup was therefore suitable to power some of the wearable devices shown in Table 7.1 (Zhao and You 2014).

Another form of wearable is the backpack, which can harvest electrical energy by walking (Granstrom et al. 2007). Compared with a traditional backpack, power harvesting backpacks replaced traditional straps with those made from PVDF piezoelectric materials. Thus, as the wearer walks with the backpack, the differential forces between wearer and the backpack would be transferred to the polymer straps, which convert the applied force to electrical energy. Such systems could generate approximately 46 mW of power using two 52 μm thick piezoelectric straps via a 100 lb load.

Body joints are also attractive places for power harvesting purposes, due to their high motion amplitude, fast angular velocity, large impulse force, and high frequency of use in human daily activities (De Pasquale, Soma, and IEEE 2013). For example, the knee-joint flexes and extends once per second during walking. The total rotation angle of the knee is approximately 70° in each direction. Michele developed a knee-joint wearable device that harvested up to 17 mW of power for a speed of 60 rpm. The device applied plucking-based frequency up-conversion to transfer low-frequency excitation input of the knee joint into high-frequency vibration of the piezoelectric bimorphs (Pozzi and Zhu 2012, 2011). The knee joint harvester was worn on the external side of the knee and was fixed by braces. The internal hub of the device with a number of plucked bimorphs rotated as the knee bent and extended during normal gait. The bimorph plucked the multiple plectra on the outer ring, which enabled it to generate high-frequency vibrations to achieve the high efficiency in converting elastic energy into electrical energy.

Moreover, another curved piezoelectric generator was recommended for wearable applications, which could also harvest electrical energy at low frequencies from body joints (Jung et al. 2015). The whole application consisted of two curved piezoelectric generators connected back-to-back. Each generator comprised a curved PI substrate and two PVDF piezoelectric films placed in a sandwich structure. It can be attached to a watch strap for harvesting power from various wrist motions. Most motions can generate output voltages exceeding 5 V and output currents exceeding 5 μ A. The generator was designed to extend the operation time of the smart watch in the future.

Furthermore, Antonino and his colleagues in 2016 have investigated different body joints to determine the best location for maximum power harvesting purposes Proto et al. (2016). In their exercise, seven body joints (neck, shoulder, elbow, wrist, hip, knee, and ankle) were compared with 12 different resistive loads by both PZT and PVDF piezoelectric materials. The results showed that the elbow joint generates the highest output power, which is approximately 10.02 nW during arm bending.

As previously discussed, piezoelectric generators transfer vibrational human body movements to electrical energy using wearable products such as backpacks, insoles, or large mechanisms attached to the human body. However, some of these wearable products are large in size and may hinder the wearer's normal gait. In this case, implantable piezoelectric harvesters have been proposed, and they are mainly used in biomedical device applications. For example, the motion of internal organs such as the heart, lungs, and diaphragm provide inexhaustible sources of energy during a human's lifespan. This provides a promising opportunity to power wearable devices.

Canan's work focused on developing advanced materials for extracting energy from natural contractile and relaxation motions of organs, which would enable high-efficiency mechanical-to-electrical energy conversion (Dagdeviren et al. 2014). The fundamental capacitor structure was a PZT layer that was sandwiched between a TI/Pt bottom layer and a Cr/Au top layer. A flexible PZT mechanical energy harvester was designed. Their device was encapsulated using a polyimide biocompatible material, which isolated the piezoelectric elements from bodily fluids and tissue, thereby minimizing any potential adverse side effects. The largest in vitro time-averaged output power of the flexible PZT device was 1.2 $\mu W/cm^2$, which was demonstrated on a bovine heart when using multilayer stacks.

Blood pressure is an alternative method of continuous energy harvesting. As Starner mentioned in his article (Starner 1996), the power generated by an average blood flow is 0.93 W and this number can be easily doubled when running. An in vivo piezoelectric generator was therefore designed using a square PZT-5A thin plate, which was placed on one side to the blood pressure, and on the other side to some constant pressure chambers. The simulated output power of the piezoelectric generator was in the micro watt range continuously, and reached a few mW intermittently. This enables the powering of MEMS devices.

Therefore, piezoelectric energy harvesters are environmentally friendly and renewable. They could supply continuous energy regardless of the surrounding environment. Piezoelectric harvesters could be designed to be flexible and small to meet human body vibration requirements. Therefore, they are easily attached on shoes, backpacks, body joints, or even internal organs. However, PZT is still the most popular piezoelectric material in recent devices, even though it is brittle and toxic. Thus, more research is needed into advanced materials that are more flexible and lead-free. Another limitation of piezoelectric transducers is that they provide AC voltage. Thus, an interface circuit is required to convert AC to DC electricity for energy storage or operating electronic devices. This process increases system complexity and cost and reduces the overall system efficiency.

7.2.3 Thermal Energy Harvesting

Thermoelectric devices convert thermal energy into electricity via the Seebeck effect. Such thin, flexible, and portable devices are capable of generating a sustainable amount of power for wearable applications Ghomian and Mehraeen (2019). The purpose of this section is to highlight how thermoelectric energy conversion is achieved and the latest trends in materials research in this area.

The Seebeck effect is a thermoelectric phenomenon that involves converting a temperature difference into a voltage difference. This phenomenon mainly occurs in metals and semiconductors. Heating one end of a semiconductor causes a temperature difference, which enables carriers to diffuse from the hot to the cold ends of the semiconductor. Considering a p-type semiconductor as an example, due to the high concentration of holes, the majority of carriers will diffuse from the heated side to the cooler side in the semiconductor. The minority carriers (electrons) will move in the opposite direction. In open circuit conditions, negative charges at the hot end and positive charges at the cold end are formed at either side of the semiconductor, which results in an electric field to appear inside the semiconductor. When the semiconductor reaches a stable state, the electromotive force caused by this temperature difference appears at both ends of the semiconductor. Both p and n type semiconducting materials are required to cause current flow in a thermoelectric generator, as shown in Figure 7.7.

Since the principle of thermoelectric generators (TEG) is based on a temperature difference, converting thermal energy into electric energy means that two major steps for TEGs are required to power wearable devices. Firstly, harvest heat energy from the user's wrist and the environment. Secondly, find an efficient way to amplify the electric energy that is generated by the TEG so that it can power wearable wristbands.

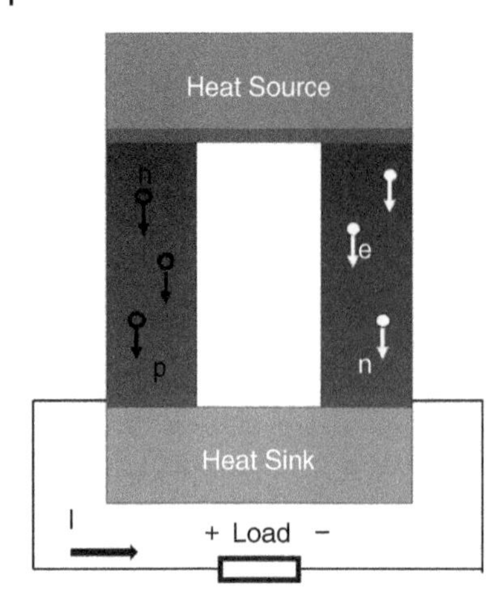

Figure 7.7 Thermoelectric Generator Made of n-type and p-type Semiconductor Thermoelectric Materials. Image adapted from Ghomian and Mehraeen (2019).

Thermoelectric efficiency of materials can be evaluated by thermoelectric figure of merit (ZT), and the definition of ZT is shown via equation (7.3)

$$ZT = S^2 T\sigma/K \tag{7.3}$$

where S is the thermoelectric power or Seebeck coefficient, T is the absolute temperature, σ is the conductivity and K is the coefficient of thermal conductivity. From equation (7.3), it is clear that there are two ways for increasing the magnitude of ZT, which involve (i) increasing $S^2\sigma$ (insulators have higher S and lower σ while metals have the opposite properties) or (ii) decreasing K.

It is worth mentioning that $S^2\sigma$ includes the following four parameters: the scattering parameter, the density of states, mobility, and the Fermi level.

The first three items are generally regarded as the intrinsic properties of materials, which can only be improved by better and purer materials. At present, the Fermi level can only be changed by adjusting the concentration of carriers. Also, since resistivity is inversely proportional to the product of carrier concentration and mobility, the lower resistivity (the higher conductivity), the better of the material.

There are several different materials that can be used as TE material. These can be divided into three broad groups:

1. Semiconductors: Semiconductors have been widely used in TEGs for their high Seebeck coefficients (more than $100\mu V/°C$). Semiconductors like Bi_2Te_3, Sb_2Te_3 have been widely used in TEGs due to their low cost and appropriately strong thermoelectric effect. For example, they have been widely used in energy harvesting, chip cooling, and chip sensing (Mamur et al., 2018).
2. Ceramics: Metal oxides have higher chemical stability, oxidation resistance, less toxicity, and lower cost in comparison to TE alloys. Therefore, they can be used in wearable devices that require a higher degree of durability (Pourkiaei et al. 2019). Moreover, ceramics are known as a noteworthy TE material for thermoelectric energy conversion and usage of them allows thermal recovery applications in combustion engines or incinerators. However, owing to the low carrier mobility of ceramic material, they were

not considered an ideal choice for TE applications material before the development of $Na_xCo_2O_4$ oxides (Pourkiaei et al. 2019).

3. Polymers: Some of the most important disadvantages of inorganic thermoelectric materials are their toxicity, limitation of natural resources, and complex expensive production procedures. So, we need to find new materials that are eco-friendlier (Pourkiaei et al. 2019) and more environmentally benign in comparison to other conventional thermoelectric modules. Mechanical flexibility, low-cost fabrication, solution process facility, and light weight are some important characteristics of the polymers that scientists have been working on. For example, some scientists are interested in the feasibility of utilizing polymers in the thermoelectric devices that were powered based on the heat of the human body (Elmoughni et al. 2019) while others have found that thermoelectric properties of composites were significantly increased by adding tellurium particles (13 vol%), bismuth telluride particles (2 vol%) and carbon black (2 vol%) (Han and Chung 2013).

Based on the above, semiconducting materials (especially Bi_2Te_3) are among the most promising for thermoelectric generation purposes (Sedky et al. 2009). Unlike ZnO, which can handle high temperatures, Bi_2Te_3 can work in room temperature applications. The most important part is relatively high ZT values. The amazing effect of Bi_2Te_3 is based on the quantum spin Hall effect. Quantum spin Hall effect can cause the electrons to flow without resistance, which means when a voltage is applied to a topological insulator, the special spin current will flow without heating and dissipation.

Currently, materials mostly made of Bi-Te are listed in Table 7.3. Assuming that room temperature is 293K, it is easy to calculate ZT at room temperature, and we can see from the table that the p-type Bi2Te3/Sb2Te3 yields the highest ZT of 2.34 at 293K. From Table 7.3, it is clear that the value of ZT is relatively low for ceramics, polymers, and some other semiconductor materials comparing to Bi-Te materials at room temperature (Hu et al. 2014). Despite the small value of ZT, the other features, such as high-temperature resistance and flexibility, make them a trustworthy alternative for TE material. Consequently, polymers and ceramics can be properly used in an alternative way (Pourkiaei et al. 2019).

7.2.3.1 Latest Trends

To power some of the wearable devices mentioned in Table 7.1, research is underway to increase the ZT and the thermoelectric efficiency of TEGs. These efforts can be grouped into three broad categories:

1. Doped material: Generally speaking, the mechanical properties of thermoelectric materials are poor. Taking Bi_2Te_3 as an example, the structure of the material is -Te-Bi-Te-Te-Bi-Te-' layered structure, and the binding structure between Te and Te is van der Waals bonding, which is easy to break. Therefore, the Te-Te layer is easy to slip when receiving pressure, leading to break and deformation. This greatly reduces the service life and range of the material. So, due to the intrinsic brittleness of Bi_2Te_3 thin films, infiltrating poly(3,4-ethylenedioxythiophene) polystyrene sulfonate (PEDOT: PSS) onto Bi_2Te_3 surfaces can improve flexibility (Kong et al. 2019). However, this method will degrade thermoelectric performance.

Table 7.3 The Value of Figure of Merit (ZT) of the Bi-Te-Based Material and the Other Common TE Material (Pourkiaei et al. 2019)

Material	Figure of merit	Temperature (k)	ZT at room temperature
$Bi_2Se_0.5Te_2.5$	1.28	298	1.26
$(Bi, Sb)_2Te_3$	1.41	298	1.39
$Bi_2Te_2.7Se_0.3$	1.27	298	1.25
$Bi_0.4Sb_1.6Te_3$	1.26	298	1.24
$p-type(Bi, Sb)_2Te_3$ thermoelectric material	1.17	323	1.06
$Bi_2(Te, Se)_3$	1.01	298	0.99
P-type$(Bi_0.26Sb_0.74)_2Te_3$+3%Te ingots	1.12	298	1.10
Bi-Sb-Te materials	1.15	350	0.96
$(Bi_2Te_3)_0.25(Sb_2Te_3)_0.75$	1.80	723	0.73
$Bi_2Te_2.85Se_0.15$	2.38	773	0.90
$Bi_0.5Sb_1.5Te_3$	1.93	693	0.82
$Bi_2Te_3-Sb_2Te_3$	1.26	420	0.88
$95\%Bi_2Te_3-5\%Bi_2Se_3$	1.67	723	0.68
$90\%Bi_2Te_3-5\%Sb_2Te_3-5\%Sb_2Se_3$	1.77	693	0.75
Bi_2Te_3	1.62	693	0.68
$Bi_2Te_2.85Se_0.15$	1.86	693	0.79
p typeBi_2Te_3/Sb_2Te_3	2.4	3 00	2.34
$Si_0.8Ge_0.2$	0.66	1073	0.18
$BaUO_3$	1.8	900	0.59
SiC/B4C+PSS	1.75	873	0.59
$Fe_0.9Mn_0.1Si_2$	1.31	773	0.51
Zn_4Sb_3	1.4	670	0.54
Tl_9BiTe_6	0.86	590	0.43
$Cu_xSn_1S_4$	0.6	570	0.31
Zn_4Sb_3	1.2	460	0.76
Graphite	0.54	393	0.41
PbTe	0.87	293	0.87
3,4-Ethylenedioxythiophene	0.42	298	0.41

Another example (Chanprateep and Ruttanapun 2018) is about doped Al_2O_3 into ZnO to improve thermoelectric properties. Since Bi_2Te_3 has low melting point, ZnO is used in high temperature applications. However, the ZnO has displayed the poor ZT value of thermoelectric properties. So, by doped with Al2O3 as formula $Zn_{1-x}Al_xO$, we can both improve thermoelectric properties of ZnO and operate at high temperature.

2. Multilayer: This type of generator is composed of several layers and is connected electrically in series or in parallel. Owing to this structure, heat flows parallel to the surface of p- and n-type films, and it is possible to easily achieve a higher temperature difference between the hot and cold sides. (Takayama and Takashiri, 2017)
3. Nanotechnology/quantum physics: By using new materials to break through technical problems in principle. For example, nano structuring has been used to decrease heat conductivity K, because a material with an all-scale hierarchical architecture (mesoscale, nanoscale, and atomic scale) would scatter phonons on each length scale (Bittner et al. 2019).

Consequently, due to the human body's temperature, this heat energy can be effectively exploited for powering emerging wearable and implantable medical devices. This range of items that can be realized includes wearable ECGs, health monitors and fitness trackers (Jaziri et al. 2019). Moreover, owing to the relatively high ZT value at room temperature, most TE generators are based on Bi_2Te_3. As for wearable devices, thermoelectric generators have many advantages that include light weight and flexibility. However, these energy harvesters are not ideal in environments that have a similar temperature to the human body. Moreover, since TEGs have polarity, they may need to be flipped when the temperature difference on both sides is negative.

7.2.4 RF Energy Harvesting

Radio frequency energy harvesting is a type of wireless power transmission system. It involves transmitting electricity from one point to another wirelessly, without the aid of cables (Brown, 1996). The history of RF energy harvesting dates back to 1969, when William C. Brown used microwaves to power helicopters (Brown 1969). Nowadays, RF energy harvesting refers to the process of using antennas and circuitry to collect and convert radio energy in the environment into DC power.

RF energy transfer uses radio frequencies ranging from 300GHz to as low as 3kHz (Lu et al. 2015) and it can be considered as a far field energy transfer technique. Therefore, RF energy transfer is suitable for powering a large number of devices distributed over a large area. An RF energy harvesting system usually consists of an antenna, an impedance matching network, a rectifier/voltage doubler, and a power management chip. Figure 7.8 shows a typical block diagram of an RF energy harvesting system.

One of the greatest benefits of this technology is that it provides a promising solution for replacing batteries in low-power equipment and systems. As of today, most low-power embedded systems and remote sensing devices use batteries as their main energy source. However, as previously mentioned, these batteries can be difficult to charge if they are distributed across a large number of nodes (Shaikh and Zeadally 2016). Secondly, harsh weather conditions and thermal cycling can damage the battery, causing chemical leaks that have adverse risks to human health (Tiliute 2007). Thirdly, all batteries will leak some current, regardless of whether they are connected to a load or not (Shaikh and Zeadally 2016). Finally, all batteries have a finite lifetime, which means that eventual replacement is inevitable (Vullers et al. 2009a).

Consequently, there is a need to develop energy harvesting techniques that can extend the batteries' lifetime, or replace them altogether. Using energy harvesting technology,

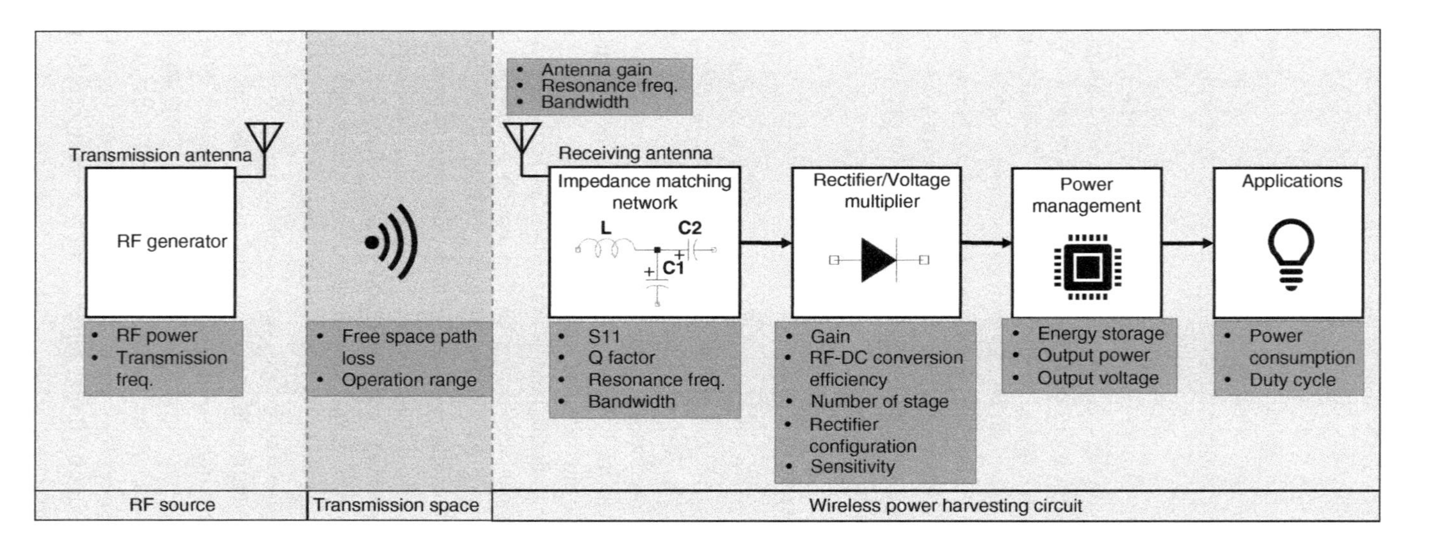

Figure 7.8 Block Diagram of an RF Energy Harvesting System. *Source*: Tran et al. (2017 Fig.1).

wearable electronic devices would become autonomously operated. This enables the devices to monitor vital human signs without interruption. Comprehensive reviews of different energy harvesting methods for low-power devices are available in the literature (see Akhtar and Rehmani 2015; Penella and Gasulla 2007; Shaikh and Zeadally 2016; Vullers et al. 2009a). As previously mentioned, the most common energy harvesting sources are solar (see Abdin et al. 2013; Brunelli et al. 2008; Raghunathan et al. 2005), wind (see Ackermann 2000; Herbert et al. 2007; SAHIN 2004), thermal (see Cuadras, Gasulla, and Ferrari 2010, Dalola, Ferrari, and Marioli 2010; Lu and Yang 2010), kinetic (see Beeby, Tudor, and White 2006; Beeby et al. 2007; Challa et al. 2008; Khaligh, Zeng, and Zheng 2010), and electromagnetic energy (see Cao et al. 2007; Tran et al. 2017; Yang et al. 2009).

In comparison to thermal or kinetic energy, electromagnetic energy is not limited by space or time. The RF collection device can be operated around the clock in urban or rural areas. The low density of RF energy is a disadvantage, but we can still meet the needs of the load by building a boost circuit (Tran et al. 2017). These characteristics can meet the needs of the Internet of Things and wireless sensor networks, which is why this technology has recently been seen as a breakthrough, and a viable next-generation powering method in those areas (Lu et al. 2015; Xie et al. 2013). In this regard, there have been many reviews from other researchers for reference. In (Shaikh and Zeadally 2016), the authors reviewed the RF energy harvesting technology and system architecture used in wireless sensor networks, including RF to DC conversion circuits and antenna design solutions. In Sudevalayam and Kulkarni (2011), the authors introduce several energy sources of sensor networks, and also include a simple application of RF energy harvesting technology and a brief introduction to RFID technology. In Lu et al. (2015), the author comprehensively introduces wireless networks with RF energy harvesting capabilities, including energy propagation models, existing applications for RF energy harvesting, circuit design, and related communication methods. It is worth noting that due to different principles, the energy harvesting efficiency of the system with the highest communication efficiency will not be maximized.

As related requirements continue to grow, commercial products based on RF energy transfer technologies such as Powercast and Cota systems have been introduced on the market. Related information can be viewed on their respective websites (Corp. 2019; Inc. 2019). At the same time, some researchers have conducted research on such commercial RF energy harvesting equipment. In Baroudi et al. (2012) and Nintanavongsa et al. (2012), the authors tested and simulated the Powercat P1100 chip and P2110-EVAL-01 development kit, verifying its performance in terms of power and efficiency.

Now, in order for RF energy harvesting technology to be further applied in the field of wireless sensor networks, wearable devices and the Internet of Things, we need to further increase system integration, reduce its volume, and increase its efficiency.

Understanding the characteristics of electromagnetic waves is essential for building an RF energy harvesting system. The performance of electromagnetic waves will vary according to distance from the transmitting antenna, frequency of the transmitting media, and conduction environment. In order to facilitate the research, according to the different characteristics of electromagnetic waves at different distances, we divide them into near-field electromagnetic waves and far-field electromagnetic waves (Yaghjian 1986). The near-field area is considered as space within the Fraunhofer distance, while the far-field area is outside the Fraunhofer distance. Fraunhofer's distance is defined as:

$$d_f = \frac{2D^2}{\lambda} \tag{7.4}$$

where d_f is the Fraunhofer distance, D is the maximum size of the radiator (or the diameter of the antenna), and λ is the wavelength of the electromagnetic wave. Although Fraunhofer created the border areas, the actual transition between the areas was not obvious. In the near field, the electrical and magnetic components are very strong and independent, so one component can dominate the other. Electric and magnetic fields behave abnormally in the near field. The relationship between electromagnetic waves in the near field varies with time and space and is difficult to predict. Thus, it is difficult to estimate the power density of this range (Tran et al. 2017). Moreover, we call the space between the antenna and a distance of $\sqrt[3]{\frac{D}{2\lambda}}$, a non-radiative/reactive near-field region, that is because E and H fields in that region are not in phase, which could create energy distortion. However, the far-field electromagnetic wave pattern is relatively uniform. For transmitter-receiver systems in far-field free space, the harvested RF power can be described by the Friis equation (Balanis 2016):

$$P_R = P_T \frac{G_T G_R \lambda^2}{(4\pi d)^2 L} \tag{7.5}$$

where P_R is the received power, P_T is the transmit power, L is the path loss factor, G_T is the transmit antenna gain, G_R is the receive antenna gain, λ is the wavelength emitted, and d is the distance between the transmit antenna and the receiver antenna. By using the path loss equation, the signal power in the far-field region can be calculated. However, this is a deterministic model. The equation does not provide all the factors that affect the propagation process, such as reflection, diffraction, absorption, and so on. In contrast, probabilistic models can take these factors into account, resulting in a more accurate model. Among them, the Rayleigh model is a probability model that is widely used in engineering practice (Rappaport et al. 1996). This model can be used to describe the situation where there is no line-in-sight channel between the transmitter and the receiver. Received power in the Rayleigh model can be calculated by equation (7.6):

$$P_R = P_R^{det} \times 10^L \times |r|^2 \tag{7.6}$$

where P_R^{det} is the received RF power calculated by a deterministic model. The path loss factor L is defined as $L = -\alpha \log 10(d/d_0)$, where d_0 is a reference distance, r denotes a random number following complex Gaussian distribution. In addition to these two common models, in Sarkar et al. (2003), the author also introduced in detail more RF propagation models in different environments.

In modern society, electromagnetic fields are used in all aspects of life, especially communications, broadcasting, medical, civilian electronic equipment, and military. Moreover, even electromagnetic waves belonging to non-ionizing radiation may be absorbed by the human body, thereby generating a thermal effect on the tissue. Although there is no consistent report confirming that long-term exposure to the radio frequency will have a serious impact on human health, for safety reasons it is still necessary to limit the RF transmission power. Therefore, each country has strict regulations on the use of electromagnetic waves in different frequency bands. Table 7.4 shows the limits of transmitter power in various

Table 7.4 Transmitter Power Restrictions

	Frequency band	Power	Duty cycle / Tx type	Channal spacing / BW	Region
a	2446-2454MHz	500mW EIRP 4W EIRP	Up tp 100% ≤15%	No spacing	Europe[b)]
b1	865.0-865.6MHz	100mW ERP		200kHz	Europe
b2	865.6-867.6MHz	2W ERP		200kHz	Europe
b3	867.6-868.0MHz	500mW ERP		200kHz	Europe
	902-928MHz	4W EIRP	FH(≥50 channels) or DSSS		USA & Canada
	2400-2483.5MHz	4W EIRP	FH(≥75 channels) or DSSS		USA & Canada
	2400-2483.5MHz	10mW EIRP		1MHz BW	Japan & Korea

a) EIRP stands for effective isotropic radiated power and is the transmit power multiplied by the gain of the transmit antenna. ERP is effective radiated power. EIRP is 1.64 times ERP. FH stands for frequency hopping. DSSS stands for direct sequence spread spectrum, and BW indicates band width.
b) Power levels above 500mW are restricted to use inside the boundaries of a building and the duty cycle of all transmissions shall in this case be ≤15% in any 200ms period (30ms on / 170ms off).
Source: Visser and Vullers (2013, Table 1).

countries in the 900MHz and 2.4GHz bands, which are often used for RF energy harvesting technology.

7.3 Conclusions

Numerous renewable ambient energy sources, such as solar, heat, vibration, and electromagnetic waves, exist in nature. Each ambient energy source exhibits different characteristics, and they all have both advantages and disadvantages. Table 7.5 shows the characteristics of different energy sources that can be used to replace batteries in wearable devices. In reality, when autonomous systems completely rely on ambient energy sources, the major challenges associated with harvesting a single source of energy can cause critical issues for device operation. For example, an autonomous system relying exclusively on the energy harvested by a photovoltaic panel (e.g., a solar panel) will fail in the absence of light. Instead of relying on a single source, energy harvesting of multiple sources can be complementary and enable truly autonomous operation, as mentioned in Bito et al. (2017) and Kim et al. (2014). The choice of hybrid technology highly depends on the application. For example, in wrist-worn wearable devices, a hybrid technology that relies on solar and piezoelectric materials might be best, so as to harvest electricity from light and the wearer's movements. Alternatively, a wearable contact lens system would favor both solar and electromagnetic energy harvesting, so that both light and RF signals can be captured.

Table 7.5 Overview of Alternative Sources of Energy to Replace Batteries

Source	Power density	Harvesting tech.	Advantages	Disadvantages
Solar	Indoor:$10\mu W/cm^2$ Outdoor:$10mW/cm^2$	Photovoltaic	High power density Mature	Not always available Required exposure to light (not implantable) Expensive
Vibration	Human:$4\mu W/cm^2$ Industrial:$100\mu W/cm^2$	Piezoelectric Electrostatic Electromagnetic	Implantable High efficiency	Not always available Material physical limitation
Thermal	Human:$30\mu W/cm^2$ Industrial:$1 - 10mW/cm^2$	Thermoelectric Pyroelectric	High power density Implantable	Not always available Material physical limitation
RF	GSM:$0.1\mu W/cm^2$ WI-FI:$1mW/cm^2$	Antenna	Always available Implantable	Low density Efficiency inversely proportional to distance

Source: Source: Tran et al. (2017, Table 1).

Thus, the low power densities from RF can be complemented with the high power density of solar energy.

Thus, since energy availability can be a temporal as well as spatial effect. To address this issue, hybrid' energy harvesting systems combine multiple harvesters on the same platform, but the design of these systems is not straightforward.

References

Abdellatif, S., K. Kirah, R. Ghannam, A.S.G. Khalil, and W. Anis (2015) Enhancing the absorption capabilities of thin-film solar cells using sandwiched light trapping structures, *Applied optics*, 54(17), p. 5534–5541.

Abdin, Z., M.A. Alim, R. Saidur, M.R. Islam, W. Rashmi, S. Mekhilef, and A. Wadi (2013) Solar energy harvesting with the application of nanotechnology, *Renewable and Sustainable Energy Reviews*, 26, p. 837–852.

Ackermann, T. (2000) Wind energy technology and current status: a review. *Renewable and Sustainable Energy Reviews*, 4(4), p. 315–374.

Adu-Manu, K., N. Adam, C. Tapparello, H. Ayatollahi, and W. Heinzelman (2018) Energy-harvesting wireless sensor networks (eh-wsns): a review. *ACM Transactions on Sensor Networks*, 14(2), p. 10:1–10:50.

Akhtar, F. and M.H. Rehmani (2015) Energy replenishment using renewable and traditional energy resources for sustainable wireless sensor networks: A review, *Renewable and Sustainable Energy Reviews*, 45, p. 769–784.

Balanis (2016) *Antenna theory: analysis and design*. Hoboken, NJ: John Wiley and Sons.

Baroudi, U., A. Qureshi, S. Mekid, and A. Bouhraoua. Radio frequency energy harvesting characterization: An experimental study (2012) *2012 IEEE 11th International Conference on Trust, Security and Privacy in Computing and Communications*, 1976–1981.

Beeby, S.P., M.J. Tudor, and N.M. White (2006) Energy harvesting vibration sources for microsystems applications, *Measurement Science and Technology*, 17(12), p. R175–R195.

Beeby, S.P., R.N. Torah, M.J. Tudor, P. Glynne-Jones, T. O'Donnell, C.R. Saha, and S. Roy (2007) A micro electromagnetic generator for vibration energy harvesting, *Journal of Micromechanics and Microengineering*, 17(7), p. 1257–1265.

Bito, J., R. Bahr, J.G. Hester, S.A. Nauroze, A. Georgiadis, and M.M. Tentzeris (2017) A novel solar and electromagnetic energy harvesting system with a 3-D printed package for energy efficient internet-of-things wireless sensors, *IEEE Transactions on Microwave Theory and Techniques*, 65(5), p. 1831–1842.

Bittner, M., N. Nikola Kanas, R. Hinterding, F. Steinbach, J. Räthel, M. Schrade, K. Wiik, M.-A. Einarsrud, and A. Feldhoff (2019) A comprehensive study on improved power materials for high-temperature thermoelectric generators, *Journal of Power Sources*, 410, p. 143–151.

Brown, W.C. (1969) Experiments involving a microwave beam to power and position a helicopter, *IEEE Transactions on Aerospace and Electronic Systems*, AES-5(5), p. 692–702.

Brown, W.C. (1996) The history of wireless power transmission, *Solar Energy*, 56(1), p. 3–21.

Brunelli, D., L. Benini, C. Moser, and L. Thiele. An efficient solar energy harvester for wireless sensor nodes (2008) *2008* Design, *Automation and Test in Europe*, (March), p. 104–109. 2008.

Cao, X., W. Chiang, Y. King, and Y. Lee (2007) Electromagnetic energy harvesting circuit with feedforward and feedback dc-dc pwm boost converter for vibration power generator system, *IEEE Transactions on Power Electronics*, 22(2), p. 679–685.

Cardwell, D., A. Kirk, C. Stender, A. Wibowo, F. Tuminello, M. Drees, R. Chan, M. Osowski, N. Pan, and Ieee (2017) Very high specific power ELO solar cells (>3 kW/kg) for UAV, space, and portable power applications, IEEE Photovoltaic Specialists Conference, p. 3511–3513. https://ieeexplore.ieee.org/document/8366552.

Challa, V.R., M.G. Prasad, Y. Shi, and F.T. Fisher (2008) A vibration energy harvesting device with bidirectional resonance frequency tunability, *Smart Materials and Structures*, 17(1), p. 015035.

Chan, R., M. Osowski, A. Wibowo, D. Cardwell, A. Kirk, C. Stender, F. Tuminello, M. Drees, N. Pan, and Ieee (2018) *High-efficiency, lightweight, flexible solar sheets with very high specific power for solar flight*, A Joint Conference of 45th Ieee Pvsc, 28th Pvsec and 34th Eu Pvsec World Conference on Photovoltaic Energy Conversion WCPEC, p. 3545–3547.

Chanprateep, S., and C. Ruttanapun (2018) Synthesis of zn0. 96al0. 04o thermoelectric material for fabrication of thermoelectric module and thermoelectric generator, *Materials Today: Proceedings*, 5(6), p. 13971–13978.

Cho, S.H., J. Lee, M.J. Lee, H.J. Kim, S.-M. Lee, and K.C. Choi (2019) Plasmonically engineered textile polymer solar cells for high-performance, wearable photovoltaics, *Acs Applied Materials and Interfaces*, 11(23), p. 20864–20872.

Cuadras, A., M. Gasulla, and V. Ferrari (2010) Thermal energy harvesting through pyroelectricity, *Sensors and Actuators A: Physical*, 158(1), p. 132–139.

Dagdeviren, C., B.D. Yang, Y. Su, P.L. Tran, P. Joe, E. Anderson, Jing Xia, et al. (2014) Conformal piezoelectric energy harvesting and storage from motions of the heart, lung, and diaphragm, *Proceedings of the National Academy of Sciences of the United States of America*, 111(5), p. 1927–1932.

Dalola, S., V. Ferrari, and D. Marioli (2010) Pyroelectric effect in PZT thick films for thermal energy harvesting in low-power sensors, *Procedia Engineering*, 5, p. 685–688.

De Pasquale, G., A. Soma, and Ieee (2013) Energy harvesting from human motion with piezo fibers for the body monitoring by mems sensors, *2013 Symposium on Design, Test, Integration and Packaging of Mems/Moems (Dtip)*, URL \<GotoISI\>://WOS:000326763300043.

Dieffenderfer, J.P., E. Beppler, T. Novak, E. Whitmire, R. Jayakumar, C. Randall, W. Qu, et al. (2014) Solar powered wrist worn acquisition system for continuous photoplethysmogram monitoring, in *2014 36th Annual International Conference of the IEEE Engineering in Medicine and Biology Society*. IEEE, p. 3142–3145.

Elmoughni, H.M., A.K. Menon, R.M.W. Wolfe, and S.K. Yee (2019) A textile-integrated polymer thermoelectric generator for body heat harvesting, *Advanced Materials Technologies*, 4(7), p. 1800708.

Ghannam R., Klaine P.V., Imran M. (2019) Artificial Intelligence for Photovoltaic Systems. In: Precup RE., Kamal T., Zulqadar Hassan S. (eds) *Solar Photovoltaic Power Plants. Power Systems*. Springer, Singapore. https://doi.org/10.1007/978-981-13-6151-7_6.

Ghomian, T., and S. Mehraeen (2019) Survey of energy scavenging for wearable and implantable devices, *Energy*, 178, p. 33–49.

Granstrom, J., J. Feenstra, H.A. Sodano, and K. Farinholt (2007) Energy harvesting from a backpack instrumented with piezoelectric shoulder straps, *Smart Materials & Structures*, 16(5), p. 1810–1820.

Green M.A. (2019) Photovoltaic technology and visions for the future. *Progress in Energy*, 1(1), p. 013001.

Green, M.A. et al. (2006) *Third generation photovoltaics*. Springer-Verlag Berlin Heidelberg. https://www.springer.com/gp/book/9783540401377

Han, S., and D.D.L. Chung (2013) Carbon fiber polymer–matrix structural composites exhibiting greatly enhanced through-thickness thermoelectric figure of merit, *Composites Part A: Applied Science and Manufacturing*, 48, p. 162–170.

Hehn, T., and Y. Manoli (2015) Cmos circuits for piezoelectric energy harvesters, *Springer Series in Advanced Microelectronics*, 38, p. 21–40.

Joselin Herbert, G.M., S. Iniyan, E. Sreevalsan, and S. Rajapandian (2007) A review of wind energy technologies, *Renewable and Sustainable Energy Reviews*, 11(6), p. 1117–1145.

Htet, K.O., R. Ghannam, Q.H. Abbasi, and H. Heidari (2018) Power management using photovoltaic cells for implantable devices, *IEEE Access*, 6, p. 42156–42164.

Hu, L.P., T.-J. Zhu, Y.-G. Wang, H.H. Xie, Z.-J. Xu, and X.-B. Zhao (2014) Shifting up the optimum figure of merit of p-type bismuth telluride-based thermoelectric materials for power generation by suppressing intrinsic conduction, *NPG Asia Materials*, 6(2), p. e88–e88.

Jaziri, J., A. Boughamoura, J. Müller, B. Mezghani, F. Tounsi, and M. Ismail (2019) A comprehensive review of thermoelectric generators: technologies and common applications, *Energy Reports*.

Jeong, E.G., Y. Jeon, S.H. Cho, and K.C. Choi (2019) Textile-based washable polymer solar cells for optoelectronic modules: toward self-powered smart clothing. *Energy & Environmental Science*, 12(6), p. 1878–1889.

Jung, W.S., M.J. Lee, M.-G. Kang, H.G. Moon, S.-J. Yoon, S.-H. Baek, and C.-Y. Kang (2015) Powerful curved piezoelectric generator for wearable applications. *Nano Energy*, 13, p. 174–181.

Kartsch, V., S. Benatti, M. Mancini, M. Magno, and L. Benini (2018) Smart wearable wristband for emg based gesture recognition powered by solar energy harvester. In *2018 IEEE International Symposium on Circuits and Systems (ISCAS)*, pages 1–5.

Kasap, S.O. (2006) *Principles of electronic materials and devices*. Tata McGraw-Hill.

Khaligh, A., P. Zeng, and C. Zheng (2010) Kinetic energy harvesting using piezoelectric and electromagnetic technologies—state of the art. *IEEE Transactions on Industrial Electronics*, 57(3), p. 850–860.

Kim, S., R. Vyas, J. Bito, K. Niotaki, A. Collado, A. Georgiadis, and M.M. Tentzeris (2014) Ambient rf energy-harvesting technologies for self-sustainable standalone wireless sensor platforms. *Proceedings of the IEEE*, 102(11), p. 1649–1666.

Kirk, A.P., D.W. Cardwell, J.D. Wood, A. Wibowo, K. Forghani, D. Rowell, N. Pan, M. Osowski, and Ieee (2018) Recent progress in epitaxial lift-off solar cells. In *7th IEEE World Conference on Photovoltaic Energy Conversion (WCPEC) / A Joint Conference of 45th IEEE PVSC / 28th PVSEC / 34th EU PVSEC*, World Conference on Photovoltaic Energy Conversion WCPEC, pages 0032–0035.

Kong, D., W. Zhu, Z. Guo, and Y. Deng (2019) High-performance flexible bi2te3 films based wearable thermoelectric generator for energy harvesting. *Energy*, 175, p. 292–299, 2019.

Liang, X., R. Ghannam, and H. Heidari (2018) Wrist-worn gesture sensing with wearable intelligence. *IEEE Sensors Journal*, 19(3), p. 1082–1090.

Liang, X., H. Li, W. Wang, Y. Liu, R. Ghannam, F. Fioranelli, and H. Heidari (2019) Fusion of wearable and contactless sensors for intelligent gesture recognition. *Advanced Intelligent Systems*, 1(7), p. 1900088.

Liu, J., Y. Li, S. Arumugam, J. Tudor, and S. Beeby (2018a) Screen printed dye-sensitized solar cells (dsscs) on woven polyester cotton fabric for wearable energy harvesting applications. *Materials Today-Proceedings*, 5(5), p. 13753–13758, 2018a.

Liu, J., Y. Li, M. Li, S. Arumugam, and S.P. Beeby (2019a) Processing of printed dye sensitized solar cells on woven textiles. *Ieee Journal of Photovoltaics*, 9 (4), p. 1020–1024, 2019a.

Liu, J., Y. Li, S. Yong, S. Arumugam, and S. Beeby (2019b) Flexible printed monolithic-structured solid-state dye sensitized solar cells on woven glass fibre textile for wearable energy harvesting applications. *Scientific Reports*, 9.

Liu, P., Z. Gao, L. Xu, X. Shi, X. Fu, K. Li, B. Zhang, et al. (2018b) Polymer solar cell textiles with interlaced cathode and anode fibers. *Journal of Materials Chemistry A*, 6(41), p. 19947–19953.

Lu, X., P. Wang, D. Niyato, D. I. Kim, and Z. Han (2015) Wireless networks with rf energy harvesting: A contemporary survey. *IEEE Communications Surveys Tutorials*, 17(2), p. 757–789.

Magno, M., D. Brunelli, L. Sigrist, R. Andri, L. Cavigelli, A. Gomez, and L. Benini (2016a) Infinitime: Multi-sensor wearable bracelet with human body harvesting. *Sustainable Computing-Informatics & Systems*, 11, p. 38–49, 2016a.

Magno, M., G.A. Salvatore, S. Mutter, W. Farrukh, G. Troester, L. Benini, and Ieee (2016b) *Autonomous smartwatch with flexible sensors for accurate and continuous mapping of skin temperature*, pages 337–340. IEEE International Symposium on Circuits and Systems.

Magno, M., G.A. Salvatore, Petar Jokic, and Luca Benini (2019) Self-sustainable smart ring for long-term monitoring of blood oxygenation. *Ieee Access*, 7, p. 115400–115408.

Mamur, H., M.R.A., Bhuiyan, F. Korkmaz, and M. Nil (2018) A review on bismuth telluride (bi2te3) nanostructure for thermoelectric applications. *Renewable and Sustainable Energy Reviews*, 82, p. 4159–4169.

Michel, B., N. Moll, and R. Ghannam (2018) Light-reflecting grating structure for photovoltaic devices, May 29 2018. U.S. Patent 9,985,147.

Mueller, S., D. Wieschollek, I.J. Junger, E. Schwenzfeier-Hellkamp, and A. Ehrmann (2019) Back electrodes of dye-sensitized solar cells on textile fabrics. *Optik*, 198.

Nintanavongsa, P., U. Muncuk, D.R. Lewis, and K.R. Chowdhury (2012) Design optimization and implementation for rf energy harvesting circuits. *IEEE Journal on Emerging and Selected Topics in Circuits and Systems*, 2(1), p. 24–33.

Ossia Inc. (2019) Ossia: Proven wireless power technology you can use today, https://www.ossia.com/.

Ostfeld, A.E., and A.C. Arias. Flexible photovoltaic power systems: integration opportunities, challenges and advances. *Flexible and Printed Electronics*, 2 (1), p. 013001, 2017.

Penella, M.T., and M. Gasulla. A review of commercial energy harvesters for autonomous sensors. In *2007 IEEE Instrumentation Measurement Technology Conference IMTC 2007*, p. 1–5, May 2007.

Polman, A., M. Knight, E.C. Garnett, B. Ehrler, and W.C. Sinke. Photovoltaic materials: Present efficiencies and future challenges. *Science*, 352(6283), p. aad4424, 2016.

Polonelli, T., D. Brunelli, M. Guermandi, L. Benini, and IEEE (2018) An accurate low-cost Crackmeter with LoRaWAN communication and energy harvesting capability, IEEE International Conference on Emerging Technologies and Factory Automation-ETFA, p. 671–676.

Pourkiaei, S.M., M.H. Ahmadi, M. Sadeghzadeh, S. Moosavi, F. Pourfayaz, L. Chen, M.A.P. Yazdi, and R. Kumar (2019) Thermoelectric cooler and thermoelectric generator devices: A review of present and potential applications, modeling and materials. *Energy*, 186 (1). https://www.sciencedirect.com/science/article/abs/pii/S036054421931521X?via%3Dihub.

Powercast Corp (2019) Wireless power products, https://www.powercastco.com/.

Pozzi, M., and M. Zhu (2011) Plucked piezoelectric bimorphs for knee-joint energy harvesting: modelling and experimental validation. *Smart Materials & Structures*, 20(5).

Pozzi, M., and M. Zhu (2012) Characterization of a rotary piezoelectric energy harvester based on plucking excitation for knee-joint wearable applications. *Smart Materials and Structures*, 21(5).

Proto, A., M. Penhaker, D. Bibbo, D. Vala, S. Conforto, and M. Schmid (2016) Measurements of generated energy/electrical quantities from locomotion activities using piezoelectric wearable sensors for body motion energy harvesting. *Sensors*, 16(4).

Pu, X., W. Song, M. Liu, C. Sun, C. Du, C. Jiang, X. Huang, et al. (2016) Wearable power-textiles by integrating fabric triboelectric nanogenerators and fiber-shaped dye-sensitized solar cells. *Advanced Energy Materials*, 6(20).

Raghunathan, V., A. Kansal, J. Hsu, J. Friedman, and M. Srivastava (2005) *Design considerations for solar energy harvesting wireless embedded systems*. In *Proceedings of the 4th International Symposium on Information Processing in Sensor Networks*, IPSN '05. Piscataway, NJ: IEEE Press.

Rappaport, T.S., *Wireless communications: principles and practice*, vol. 2. Upper Saddle River, NJ: Prentice Hall. https://nyuscholars.nyu.edu/en/publications/wireless-communications-principles-and-practice-2.

Sahin, A. (2004) Progress and recent trends in wind energy. *Progress in Energy and Combustion Science*, 30(5), p. 501–543.

Sarkar, T.K., Z. Ji, K. Kim, A. Medouri, and M. Salazar-Palma. A survey of various propagation models for mobile communication. *IEEE Antennas and Propagation Magazine*, 45(3), p. 51–82.

Scheiman, D., P. Jenkins, Robert Walters, Kelly Trautz, Raymond Hoheisel, Rao Tatavarti, Ray Chan, et al. (2014) *High Efficiency Flexible Triple Junction Solar Panels.* 2014 Ieee 40th Photovoltaic Specialist Conference.

Sedky, S., A. Kamal, M. Yomn, H. Bakr, R. Ghannam, V. Leonov, and P. Fiorini (2009) Bi 2 te 3 as an active material for mems based devices fabricated at room temperature. In *TRANSDUCERS 2009-2009 International Solid-State Sensors, Actuators and Microsystems Conference*, p. 1035–1038. IEEE.

Shaikh F.K., and S. Zeadally (2016) Energy harvesting in wireless sensor networks: A comprehensive review, *Renewable and Sustainable Energy Reviews*, 55, p. 1041.

Shenck, N.S., and J.A. Paradiso (2001) Energy scavenging with shoe-mounted piezoelectrics. *Ieee Micro*, 21(3), p. 30–42.

Song, L., T. Wang, W. Jing, X. Xie, P. Du, and J. Xiong (2019) High flexibility and electrocatalytic activity mos2/tic/carbon nanofibrous film for flexible dye-sensitized solar cell based photovoltaic textile. *Materials Research Bulletin*, 118.

Starner, T. (1996) Human-powered wearable computing. *Ibm Systems Journal*, 35(3-4):618–629.

Stender, C.L., J. Adams, V. Elarde, T. Major, H. Miyamoto, M. Osowski, N. Pan, et al. (2015) *Flexible and Lightweight Epitaxial Lift-Off GaAs Multi-Junction Solar Cells for Portable Power and UAV Applications*, IEEE Photovoltaic Specialists Conference.

Sudevalayam, S., and P. Kulkarni (2011) Energy harvesting sensor nodes: survey and implications, *IEEE Communications Surveys Tutorials*, 13(3):443–461.

Takayama, K., and M. Takashiri (2017) Multi-layered-stack thermoelectric generators using p-type sb2te3 and n-type bi2te3 thin films by radio-frequency magnetron sputtering. *Vacuum*, 144, p. 164–171.

Tiliute, D.E. (2007) Battery management in wireless sensor networks, *Elektronika ir elektrotechnika*, 76(4), p. 9–12.

Tran, L.-G., H.-K. Cha, and W.T. Park (2017) RF power harvesting: a review on designing methodologies and applications, *Micro and Nano Systems Letters*, 5(1).

Trautz, K., P. Jenkins, R. Walters, D. Scheiman, R. Hoheisel, R. Tatavarti, R. Chan, et al. (2013) *High Efficiency Flexible Solar Panels*, IEEE Photovoltaic Specialists Conference, p. 115–119.

Visser, H.J., and R.J.M. Vullers (2013) RF energy harvesting and transport for wireless sensor network applications: principles and requirements, *Proceedings of the IEEE*, 101(6):1410–1423, June 2013. doi: 10.1109/JPROC.2013.2250891.

Vullers, R.J.M., R. van Schaijk, I. Doms, C. Van Hoof, and R. Mertens (2009a) Micropower energy harvesting, *Solid-State Electronics*, 53(7), p. 684–693, 2009a.

Vullers, R.J.M., R. van Schaijk, I. Doms, C. Van Hoof, and R. Mertens (2009b) Micropower energy harvesting, *Solid-State Electronics*, 53(7), p. 684–693, 2009b. ISSN 0038-1101.

Xie, L., Y. Shi, Y.T. Hou, and A. Lou. Wireless power transfer and applications to sensor networks (2013) *IEEE Wireless Communications*, 20(4), p. 140–145.

Xin, L., and S.-H. Yang (2010) Thermal energy harvesting for wsns, in *2010 IEEE International Conference on Systems, Man and Cybernetics*, p. 3045–3052, Oct 2010. doi: 10.1109/ICSMC.2010.5641673.

Yaghjian, A. (1986) An overview of near-field antenna measurements, *IEEE Transactions on Antennas and Propagation*, 34(1):30–45.

Yang, B., C. Lee, W. Xiang, J. Xie, J.H. He, R.K. Kotlanka, S.P. Low, and H. Feng (2009) Electromagnetic energy harvesting from vibrations of multiple frequencies, *Journal of Micromechanics and Microengineering*, 19(3), p. 035001.

Yuan, M., R. Das, R. Ghannam, Y. Wang, J. Reboud, R. Fromme, F. Moradi, and H. Heidari (2020) Electronic contact lens: A platform for wireless health monitoring applications, *Advanced Intelligent Systems*, 1900190. 2 (4). https://onlinelibrary.wiley.com/doi/full/10.1002/aisy.201900190.

Yun, M.J., Sim, Y.H., Cha, S.I. et al. Three-Dimensional Textile Platform for Electrochemical Devices and its Application to Dye-Sensitized Solar Cells. *Sci Rep* 9, 2322 (2019). https://doi.org/10.1038/s41598-018-38426-1.

Zhang, Z., Z. Yang, Zhongwei Wu, Guozhen Guan, Shaowu Pan, Ye Zhang, Houpu Li, et al. (2014) Weaving efficient polymer solar cell wires into flexible power textiles, *Advanced Energy Materials*, 4(11).

Zhao, J., and S. You (2014) A shoe-embedded piezoelectric energy harvester for wearable sensors, *Sensors*, 14(7), p. 12497–12510.

Zhao, J., R. Ghannam, Q.H. Abbasi, M. Imran, and H. Heidari (2018) Simulation of photovoltaic cells for implantable sensory applications, in *2018 IEEE SENSORS*. IEEE. 1–4.

Zhao, J., R. Ghannam, M.K. Law, M.A. Imran, and H. Heidari (2019) Photovoltaic power harvesting technologies in biomedical implantable devices considering the optimal location, *IEEE Journal of Electromagnetics, RF and Microwaves in Medicine and Biology*, 2019.

8

Wireless Control for Life-Critical Actions

Burak Kizilkaya[1], *Bo Chang*[2], *Guodong Zhao*[1] *and Muhammad Ali Imran*[1]

[1] *School of Engineering, University of Glasgow*
[2] *National Key Laboratory of Science and Technology on Communications, University of Electronic Science and Technology of China*

Healthcare is one of the application areas of real-time wireless communications and control. With the development of communication and control technologies, there is a potential to transfer not only observed data but also skills over wireless links. Telesurgery and remote diagnosis are examples of transferring skills with real-time wireless control. Such applications include observing patients as well as diagnosing them remotely, which is the transfer of skills of doctor to the remote location. In this chapter, we discuss real-time wireless control for life-critical actions. In particular, we introduce the basics of wireless control systems and discuss the fundamental design capabilities needed to realize real-time wireless control, with primary emphasis given to communication-control co-design. The goal is to provide integrated solutions for life-critical actions in healthcare. A co-design system model is proposed and explained in detail. Simulation results are discussed and benefits of co-design are depicted in terms of both control and communication performance.

8.1 Introduction

With rapid development of wireless network technologies, sensors, and actuators, new application areas like smart homes, smart cities, environment monitoring, and surveillance systems have started to become more and more popular. In addition, wireless communications play an important role in life-critical actions such as patient monitoring, remote diagnosis, physiological data tracking, early flood detection, early fire detection, and disaster management where sometimes the communication infrastructure is not available (Akyildiz et al. 2002; Akyildiz et al. 2007; Akyildiz and Vuran 2010).

On the other hand, according to discussion in Bello et al. (2017), most of the available solutions in the literature in the context of critical communications rely on monitoring with slow refresh rates. However, in critical applications, it is very crucial to have real-time connection over wireless networks for real-time monitoring and control. Especially in healthcare applications, monitored or controlled applications are life-critical, and the consequences of

Engineering and Technology for Healthcare, First Edition.
Edited by Muhammad Ali Imran, Rami Ghannam and Qammer H. Abbasi.

poor communication can be vital such as tissue and organ injuries, wrong diagnoses, and wrong treatments.

In near future, we are expecting wireless networks to take the place of conventional wired networks in many application areas, especially in healthcare/Potential benefits of wireless networks are ease of deployment, reduction in cost of deployment and maintenance, and flexible and adaptive systems (Araújo et al. 2014). The main problems with wireless networks are reliability and latency of the wireless links. Especially in real-time wireless control systems, the main requirements are low-latency (end-to-end delay of 1 ms or less, Cizmeci et al. 2017) and high reliability (wireless link failure probability of 10^{-7} or less (Simsek et al. 2016) over wireless links. In critical applications such parameters gain even more importance. Real-time control systems are very sensitive, and failure of the wireless link may lead to serious consequences (Park et al. 2018).

Real-time wireless control systems are not a fiction anymore, thanks to developments in the area of wireless communication and control. However, wireless control systems in life-critical actions need to be studied to bring about real-time wireless control in life-critical actions with high performance in terms of latency, reliability, and security. Today's wireless technology lags behind the requirements of real-time wireless control. For example it requires 1MS or less end-to-end latency and link failure probability of 10^{-7} or less in life critical actions. However, latency is around 20 ms and reliability is around 10^{-1} in today's communication technology of 4G (Aijaz et al. 2016).

Some serious studies have been conducted in the area of real-time wireless control for life-critical actions in healthcare. For example, the first known remote surgery was demonstrated in 2001 with constant latency of 155 ms and zero packet loss (Marescaux et al. 2002). The patient was in Strasbourg, France, and the surgeon was in New York. Another example was demonstrated in 2016 by Ericsson and King's College London using 5G technology. The most recent example of remote surgery was carried out in China in 2019. The patient was 3000 km away from the surgeon, and brain surgery was performed using 5G technology with the collaboration of Chinese PLA General Hospital (PLAGH), China Mobile, and Huawei 5G technology (ChinaDaily 2016).

Considering recent developments and emerging technologys such as the Internet of Things (IoT), the Industrial Internet of Things (IIoT), the new Tactile Internet paradigm, and the capabilities of future 5G communications, new types of sensing become available. In traditional monitoring systems, scalar data (e.g., temperature, humidity, etc.) and multimedia data (e.g., audio, video, still image, etc.) are transmitted over wireless links. On the other hand, with the tactile internet paradigm, tactile sensing (i.e., haptic/tactile feedback from environment) becomes of interest to researchers. Tactile sensing provides more realistic control especially in life-critical healthcare applications. For example, in remote surgery case, sensing the texture and viscosity of the tissues or recognizing tumors and lumps can be achieved by tactile sensing. Only audio and video streams and physiological sensor data readings (e.g. heart rate, blood pressure, etc.) are not sufficient for such scenarios.

In this chapter, real-time wireless control systems are discussed in healthcare contexts such as remote surgery and remote diagnosis. Basics of wireless control systems are introduced and fundamental design capabilities are explained, giving primary emphasis to communication control co-design. The main objective is to provide integrated solutions for life-critical actions in healthcare.

8.2 Wireless Control for Healthcare

Within the paradigm of IoT, smart cities, and smart environments, healthcare applications are receiving great interest from both the research and the industrial communities. In the beginning, scalar sensors (e.g., temperature, heart rate, blood pressure, etc.) are used in healthcare applications. Using scalar data, the patient's physiological status is observed. However, scalar data are insufficient for precise monitoring and diagnoses since the information gathered from scalar sensors is limited. To overcome limitations of scalar sensors, multimedia sensors are introduced, which are capable of transmitting multimedia data such as audio, video, and still images. In the multimedia context, more precise monitoring and early detection of vital diseases has become possible by using multimedia data in more advanced techniques such as machine learning, and image/video processing.

Nowadays, with the fifth-generation cellular network, real-time control over wireless links is receiving interest from both researchers and industry. As discussed in previous sections, applications of real-time wireless communications in healthcare are some of the main interests. Already, there are studies and successful demonstrations of real-time wireless control in healthcare such as remote surgery.

In this study, real-time wireless communication and control in healthcare applications are discussed in the context of communication control co-design. A real-time wireless control system consists of three main components: i) master robot, ii) slave robot, and iii) communication network as shown in Figure 8.1.

Master and slave robots are in physically different locations. The master robot is used by an expert (i.e., doctor/surgeon in healthcare) who is an expert in his/her area and is trained to use the robot. The master robot is equipped with controllers to control the slave robot, and audio and video output devices (e.g., screen display, speakers, etc.) to see and hear the operation scene, and video and audio input devices (e.g., camera, microphone, etc.) to be

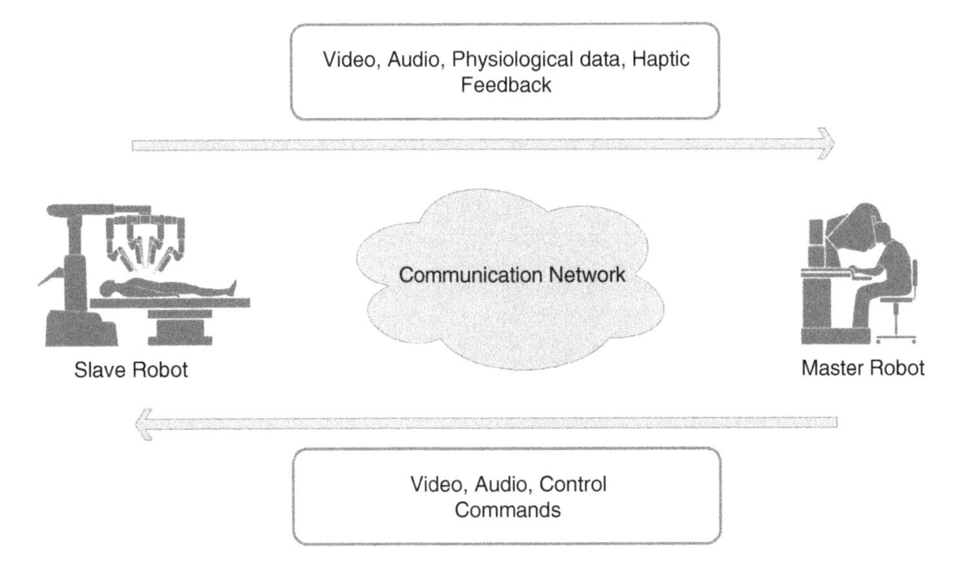

Figure 8.1 Real-Time Wireless Control System for Healthcare.

seen and heard by assistant doctors or nurses, and haptic feedback devices to sense the remote environment. The slave robot environment, on the other hand, is equipped with audio and video input devices to be seen and heard by the surgeon; audio and video output devices for the assistant doctor/nurses to see and hear the surgeon, and tactile sensors to produce haptic feedback to the surgeon. The communication network is the link to bring about communication between the master and slave robots (Zhang et al. 2018).

The system we dsicuss has various advantages. . Geographical difficulties can be overcome with wireless remote control. High quality surgeons may become available anytime, anywhere. In addition, it becomes possible to diagnose patients who are not capable of accessing healthcare services (Simsek et al. 2016). Moreover, the system decreases the risk of infection, and operations can be done with smaller operation scar which leads to faster recovery of patient.

8.3 Technical Requirements

The real-time communication and control paradigm has the potential to make real-time monitoring, diagnosis, and treatment possible. In the near future, it is expected to create a revolution in healthcare, remote surgery, remote diagnosis and remote treatment. However, considering today's technological capabilities in terms of communication, it is possible to discuss some technical requirements to realize real-time wireless control.

8.3.1 Ultra-Reliability

Reliability of the wireless link is one of the most important issues, especially in life-critical applications since wireless links are prone to failure and packet loss, which directly affect control performance. For applications such as remote surgery and remote diagnosis, a wireless link failure probability of 10^{-7} or less is expected to allow high performance control over wireless links. Ultra-reliability is important since the action itself is life-critical and serious consequences result from failure (Simsek et al. 2016). In addition, response efficiency of real-life control can be achieved with ultra-reliable communications as well as upgraded hardware to support ultra-reliability (Miao et al. 2018). Since new real-time healthcare control systems include haptic feedback, reliable communication becomes more crucial. Failures of reliability may lead to incorrect diagnoses, incorrect treatments, injuries, and even death (Aijaz et al. 2016).

8.3.2 Low Latency

To increase the stability and performance of real-time wireless control systems, another important requirement is low latency that is, end to end delay. Haptic feedback communication is generally required in every millisecond. Packet loss and high latency can create serious consequences in terms of the patient's health as well as damaging the performance of the control system (Antonakoglou et al. 2018). To ensure the stability of the system and to provide high performance, low latency over wireless transmission should be ensured. Since the current technology (i.e., LTE/4G) lags behind the requirements of such communication, the forthcoming 5G technology is expected to become the solution for such requirements where high data rates, ultra reliability, and low latency are envisioned by 5G communication technology.

8.3.3 Security and Privacy

Healthcare applications, especially remote surgery and remote diagnosis applications require high security and privacy since the life-critical actions are performed over wireless links. In this case, the security requirements of the communication systems are as important as reliability and low latency. Real-time wireless communication and control systems can face some malicious actions such as DoS attacks by UDP or TC flooding, malicious code injections into the application by a buffer overflow, altering transmitted packets between the master and slave robot illegitimately, and replaying some legitimate packets (El Kalam et al. 2015). Such malicious actions can lead to serious problems for the health of the patient and the performance of the system. With new real-time control and communication systems, transmitted data become more vital since the life-critical actions are being performed. To overcome security issues, new network security approaches should be investigated by collaborating with developing 5G technology. In addition, the tradeoff between security of the system and latency requirements should be investigated to ensure that new security schemes meet with both security and QoS requirements (Antonakoglou et al. 2018).

8.3.4 Edge Artificial Intelligence

Real-time communication and control systems increase the number of technological devices in many areas, as well as in healthcare. In addition, edge computing techniques are popular for decreasing the use of resources since computing activity is performed close to the site where data are generated (i.e., on the edge). To overcome latency, reliability, or packet loss problems, artificial intelligence (AI) solutions can be deployed to the edge of the network to realize predictive control and caching. For example, in the presence of packet loss, predictive packets can be used from the buffer to keep the performance of the system stable in real-time control systems (Simsek et al. 2016). It is becoming inevitable that scientists will research and discuss new machine learning and AI techniques to use wireless network resources optimally as well as to overcome problems of wireless communications such as packet loss and latency.

Technical requirements of real-time communication and control in healthcare applications can be summarized in terms of transmitted data types as shown in Table 8.1.

8.4 Design Aspects

Real-time wireless control systems consist of sensors (all scalar, multimedia, and haptic), actuators, and controllers that are controlled through wireless links. In the new real-time wireless control era, wireless links are expected to replace traditional wired communications. In this sense, the main challenge become the design of new paradigm where control

Table 8.1 Technical Requirements of Real-Time Wireless Communication and Control in Healthcare applications

	Latency	Packet loss rate	Reference
Scalar Data (heart rate, blood pressure, temperature, respiration rate)	< 250 ms	$< 10^{-3}$	Patel and Wang (2010)
Multimedia data (audio, video)	< 150 ms	$< 10^{-3}$	Perez et al. (2016); Eid et al. (2010); Cizmeci et al. (2017)
Haptic feedback data	< 3 ms	$< 10^{-4}$	Eid et al. (2010); Cizmeci et al. (2017); Hachisu and Kajimoto (2016)

Table 8.2 Independent Design vs Co-Design

	Independent design	Co-design
Complexity	Less complex, easier to model	More complex and relatively difficult models
Overall performance	High control performance but low overall performance	High overall performance since parameters are jointly tuned
Wireless resource usage	Low optimization of wireless resources since control requirements are considered first	High optimization of wireless resources since both wireless resources and control parameters are jointly optimized considering tradeoff between them
Flexibility	More flexible control systems since they are suitable with different communication systems	Flexibility is less since two systems are modeled as one system
Failure detection/diagnosis	Failure detection and diagnosis are relatively easy since the complexity of the system is low	It's challenging to detect and diagnose failures because of system's complexity

and communication systems are tightly coupled and interacted more than before. There are two main design approaches; *i)independent/separate design, ii)joint/co-design* (Park et al. 2018). In this section, two design approaches for real-time wireless communication and control is discussed. The benefits of co-design is highlighted along with challenges. In addition, comparison of two approaches are conducted as summarized in Table 8.2.

8.4.1 Independent Design

In the independent design approach, control and communication systems are designed separately. Generally, control systems are designed first. Then, according to requirements of the control systems in terms of communication such as data rate, latency, and reliability, the communication system is designed to fulfill the requirements specified (Zhao et al. 2018). In other words, communication parameters are tuned to fulfill control system requirements.

This approach simplifies the system design as well as creating flexibility, so that control systems are generally compatible with different communication systems. However, the main problem with independent design is that control requirements are much more important than the performance and QoS requirements. It creates the risk of vast use of wireless resources (Park et al. 2018). In brief, independent design uses communication as a tool to fulfill the requirements of control, yet it does not consider communication requirements as much as control requirements. As a result, the system that is created fulfills the control requirements but does not operate at the optimum level in terms of performance, resources (e.g., wireless resources, and energy resources) management, and QoS.

8.4.2 Co-Design

In the co-design approach, both control and communication parameters are tuned jointly to achieve better overall performance. Instead of considering two separate systems, the co-design approach considers the system as a whole and optimizes the performance parameters jointly by considering the tradeoff between communication and control system parameters (Park et al. 2018). In real life, the system performance relies not only on control performance but also on the performance of the communication system (Zhao et al. 2018). In such a case, only fulfilling the requirements of the control system does not leverage the performance of the whole system to higher levels. In independent design, it is more flexible and less complex to design the systems where it is more challenging in co-design. However, considering two separate systems degrades overall system performance as well as consuming vast amounts of wireless resources. In short, the main goal of the co-design approach is to consider both control and communication parameters jointly to design better system in terms of overall system performance since the overall performance of the system depends not only on control performance but also on communication performance. In addition to ensuring proper communication, resource management of the wireless links is also crucial to satisfy the latency and packet loss requirements of real-time control.

8.5 Co-Design System Model

A typical real-time wireless control system as shown in Figure 8.2, can be seperated as uplink and downlink. The sampling signal that shows the plant state is transmitted from the sensor to the remote controller through the uplink. Then, the remote controller transmits the control input signal to the plant by downlink. The downlink is assumed to be perfect, but transmission delay and packet loss occur in the uplink. In this system, communication time delay and packet loss are taken into account to obtain real-time wireless control functions. To evaluate system performance, two metrics are discussed in terms of communication and control respectively (Chang et al. 2019).

8.5.1 Control Function

Plant states are transmitted to the remote controller by the sensor. Then, the remote controller calculates the control input and transmits to the plant. Lastly, the plant state is updated in light of the current state and the control input. Considering the mentioned process and as discussed in Park et al. (2011), the continuous control system is given by a

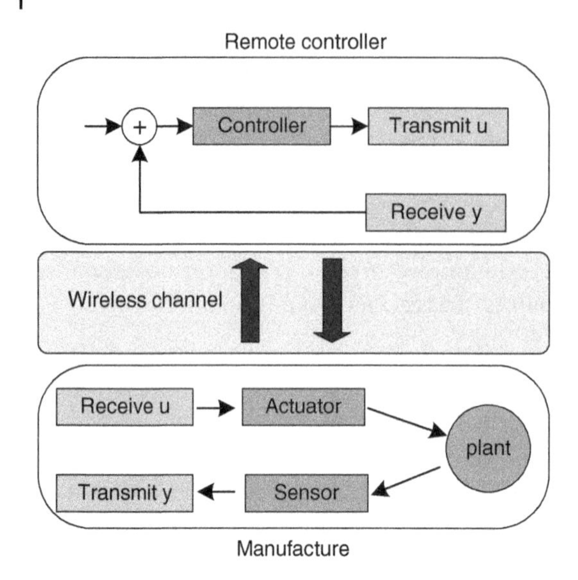

Figure 8.2 Real-Time Wireless Control System. Source: Chang, B., Zhao, G., Zhang, L., Imran, M.A., Chen, Z., and Li, L. 2019 (© [2019] IEEE).

linear differential equation

$$d\mathbf{x}(t) = \mathbf{A}\mathbf{x}(t)dt + \mathbf{B}\mathbf{u}(t)dt + d\mathbf{n}(t), \tag{8.1}$$

where $\mathbf{x}(t)$ is the plant state at time t, $\mathbf{u}(t)$ is the control input, and $\mathbf{n}(t)$ is the process disturbance caused by additive white gaussian noise (AWGN) with zero mean and variance $\mathbf{R}_n$. Additionally, matrices $\mathbf{A}$ and $\mathbf{B}$ represent parameters of the control system, and they have real physical meanings. Further details can be found in Cai and Lau (2018) and Boubaker and Iriarte (2017).

To achieve the discrete time control system, we assume h_k represents the time varying sample period at the sensor. It consists of the wireless transmission time delay d_k and a constant idle period $\overline{d}$. It is possible to express their relationship as

$$h_k = \overline{d} + d_k, \tag{8.2}$$

where the sample period h_k is affected by the transmission time delay d_k. The time index of the time sequential control process is depicted as $k = 1, 2, 3, \cdots, N$. Therefore, the discrete time control model considering transmission time delay d_k can be expressed as

$$\mathbf{x}_{k+1} = \mathbf{\Omega}_k\mathbf{x}_k + \mathbf{\Phi}^{k_0}u_k + \mathbf{\Phi}^{k_1}u_{k-1} + \mathbf{n}_k, \tag{8.3}$$

with observations at the sensor

$$\mathbf{y}_k = \mathbf{C}\mathbf{x}_k + \mathbf{n}'_k, \tag{8.4}$$

where $\mathbf{\Omega}_k = e^{\mathbf{A}h_k}$, $\mathbf{\Phi}^{k_0} = \left(\int_0^{\overline{d}} e^{\mathbf{A}t}dt\right)\mathbf{B}$, $\mathbf{\Phi}_1^k = \left(\int_{\overline{d}}^{h_k} e^{\mathbf{A}t}dt\right)\mathbf{B}$, $\mathbf{n}'_k \in \mathbb{R}^n$ is the disturbance caused by AWGN with zero mean and variance $\mathbf{R}'_n$, and $k = 0, 1, 2, \cdots, N$ is the sample time index in the control process.

The probability of packet loss is depicted as ε. Then, $\Pr\{l_k = 1\} = 1 - \varepsilon$ becomes the probability of successful transmission at time index k. Therefore, we can express the samples that are received by the controller as

$$\hat{\mathbf{y}}_k = \begin{cases} \mathbf{C}\mathbf{x}_k + \mathbf{n}'_k, & l_k = 1, \\ \mathbf{0}, & l_k = 0. \end{cases} \tag{8.5}$$

By assuming the generalized state as $\xi_k = (\mathbf{x}_k^T \ u_{k-1})^T$, the control function in (8.3) and (8.4) can be rewritten as

$$\xi_{k+1} = \mathbf{\Omega}_d \xi_k + \mathbf{\Phi}_d u_k + \overline{\mathbf{n}}_k, \tag{8.6}$$

and

$$\hat{\mathbf{y}}'_k = l_k(\mathbf{C}_d \xi_k + \overline{\mathbf{n}}'_k), \tag{8.7}$$

where $\overline{\mathbf{n}}_k = (\mathbf{n}_k^T \ 0)$, $\overline{\mathbf{n}}'_k = (\mathbf{n}'^{T_k} \ 0)$, $\mathbf{\Phi}_d = \begin{pmatrix} \mathbf{\Phi}_0 \\ \mathbf{I} \end{pmatrix}$, and $\mathbf{C}_d = (\mathbf{C} \ 0)$. According to Schenato et al. (2007), it is assumed that $\mathbf{\Omega}_k = \mathbf{\Omega}$. Then, we have $\mathbf{\Omega}_d = \begin{pmatrix} \mathbf{\Omega} & \mathbf{\Phi}_1 \\ 0 & 0 \end{pmatrix}$.

8.5.2 Performance Evaluation Criterion

8.5.2.1 Control Performance

Control cost is generally considered as one of the performance criteria in control systems (Nilsson et al. 1998; Rasool and Nguang 2008). In this chapter, the quadratic control cost is adopted as in Schenato et al. (2007),

$$J_N = \mathbb{E}\left[\xi_N^T \mathbf{W} \xi_N + \sum_{k=0}^{N-1} (\xi_k^T \mathbf{W} \xi_k + u_k^T \mathbf{U} u_k)\right], \tag{8.8}$$

where $\mathbf{W}$ is the weight of the state and $\mathbf{U}$ is the weight of the control input. Considering the fact that the plant state has the top priority in real-time control for manufacturing, it is assumed that the weight on plant state $\mathbf{W}$ is much larger compared to the weight on the control input $\mathbf{U}$ in (8.8). Using the optimal feedback control law, the minimum J_N value is expressed as follows (Park et al. 2017):

$$\begin{aligned} J_N^* &= \xi_0^T \mathbf{S}_0 \xi_0 + \mathrm{Tr}(\mathbf{S}_0 \mathbf{P}_0) + \sum_{k=0}^{N-1} (\mathrm{Tr}((\mathbf{\Omega}_d^T \mathbf{S}_{k+1} \mathbf{\Omega}_d \\ &\quad + \mathbf{W} - \mathbf{S}_k) \mathbf{E}_{l_k}[\mathbf{P}_{k|k}]) + \mathrm{Tr}(\mathbf{S}_{k+1} \mathbf{R}_n)), \end{aligned} \tag{8.9}$$

where $\xi_0 = (\mathbf{x}_0^T, \ 0)^T$. The initial state $\mathbf{x}_0$ is a white Gaussian random variable where $\overline{\mathbf{x}}_0$ is the mean and $\mathbf{P}_0$ is the covariance. In addition, the initial value is $\mathbf{S}_N = \mathbf{W}$.

8.5.2.2 Communication Performance

Energy consumption in wireless communications is one of the communication performance criteria. In addition, occupied frequency bandwidth is assumed to be fixed. Furthermore the bandwidth is selected that provides the best channel gain. Considering the finite packet length of URLLC (Durisi et al. 2016), the channel capacity with transmission time delay and packet loss is different from the traditional Shannon capacity. Therefore, we introduce the details of the wireless resource consumption criteria.

As discussed in Durisi et al. (2016), the available uplink capacity can be expressed as

$$C_{bit} \approx \frac{T_d B_d}{\ln 2} \left\{ \ln \left(1 + \frac{h_f^2 g p}{N_0 B_d} \right) - \sqrt{\frac{V}{T_d B_d}} f_Q^{-1}(\varepsilon) \right\}. \tag{8.10}$$

where C_{bit} is the Shannon capacity where error bits introduced by channel dispersion are eliminated. Then, the channel dispersion V is expressed as

$$V = 1 - \frac{1}{[1 + \frac{h_f^2 g p}{N_0 B_d}]^2}. \tag{8.11}$$

In equations (8.10) and (8.11), T_d is the allocated time resource, B_d is the allocated frequency resource, p is the transmission power, $f_Q^{-1}(\cdot)$ is the inverse of the Q-function, ε is the error probability, N_0 is the single-sided noise spectral density, g is the path-loss, and h_f is the small-scale fading.

Without loss generality, the following channel path-loss model is considered (UTRA 2006):

$$g_{[dB]} = -128.1 - 37.6\,\lg(d), \text{ for } d \geq \delta, \tag{8.12}$$

where d shows the distance between two nodes, and $\delta = 35$ m is the minimum distance. The small-scale fading h_f follows Rayleigh distribution where the mean is zero and the variance is $\sigma_h^2 = 1$. It is assumed that the small fading channel coherence time is larger than the uplink frame duration. In other words, the block fading channel is considered (Yang et al. 2014).

In light of the discussion above, by solving (8.10) with given C_{bit} the transmission power p^* can be obtained. Therefore, we can express the energy consumption as

$$E_{tol} = \sum_{k=1}^{N} p_k^*, \tag{8.13}$$

where p^{*_k} is the transmission power for the k-th sample and the time delay is d_k, the packet loss probability is ε_k.

8.5.3 Effects of Different QoS

Figure 8.3 shows the effect of different communication QoSs on the state update. In this figure, the sampling period is 100 ms, and the initial state is (100, 100). Other simulation parameters can be found in Section 8.5.4.

From the figure, the states values for both high and low QoSs increases and then decreases with respect to k. However, changes are more rapid and sharp in low QoS as compared to high QoS.

In addition, for k between 1 and 5, the state value of high QoS is less than the state value of low QoS. Therefore, considering (8.8), we can determine that the control cost of high QoS is less than the control cost of low QoS in that period. This clearly means that high QoS outperforms low QoS in terms of control cost when $1 < k < 5$. On the other hand, for $5 \leq k < 13$, low QoS outperforms high QoS in terms of control cost. This shows that we can divide the control process into two phases before being stable. For the first phase until state value reaches the maximum, high QoS should be used to serve the control

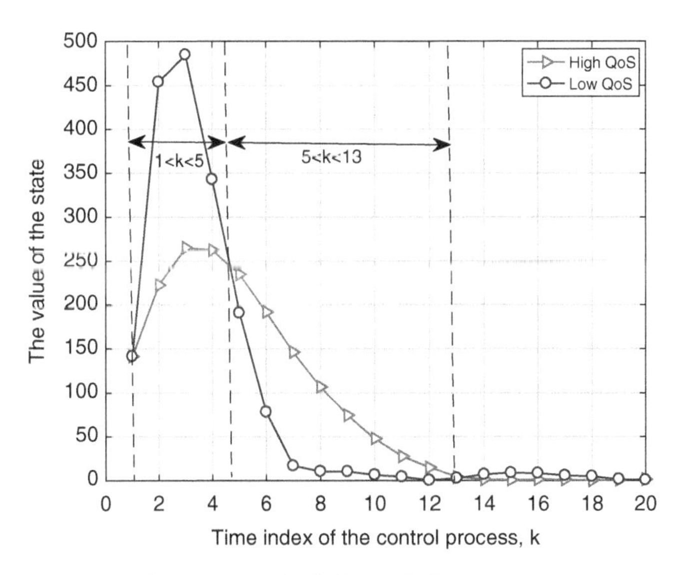

Figure 8.3 State Updates for Different QoS.

process when considering control cost. However, in the second phase, which is between largest state value and stable state value, low QoS should be used to serve control process in light of control costs as well as communication energy consumption. After achieving stability, the difference between high and low QoS becomes negligible. Therefore, low QoS should be used to serve the stable phase considering energy consumption. By applying dynamic QoS, both control cost and communication energy consumption can be maintained.

8.5.4 Numerical Results

In this section, to demonstrate the performance of co-design, following numerical results are provided. As discussed earlier, there are two types of QoS. For high QoS, 1 ms of time delay and 10^{-5} of packet loss probability is used. On the other hand, 100 ms of time delay and 10^{-3} of packet loss probability is used for low QoS (She et al. 2016). The control parameters are as follows: $\mathbf{A} = \begin{pmatrix} 2 & 14 \\ 0 & 1 \end{pmatrix}$, $\mathbf{B} = \begin{pmatrix} 0 \\ 1 \end{pmatrix}$, $\mathbf{C} = \begin{pmatrix} 1 & 0 \\ 0 & 1 \end{pmatrix}$, $\mathbf{P}_0 = 0.01\mathbf{I}$, $\mathbf{W} = \mathbf{I}$, $\mathbf{U} = 0.0001$, $\mathbf{R}_{n'} = 0.01\mathbf{I}$, and the initial state is (100,100). The length of the discrete control process is $N = 10000$. In the simulations, each packet consists of 100 bits, and the bandwidth of communication system is 1 MHz. In addition, the single-sided noise spectral density is considered as -174 dBm/Hz, and the distance between the sensor, and the controller is 100 m. To obtain each curve 10,000 Monte Carlo trails are performed. Furthermore, the control and communication performances are considered for the small disturbance noise, that is, $\mathbf{R}_n = 0.0001\mathbf{I}$.

In Figure 8.4, high QoS, low QoS, and co-design methods are compared in terms of control cost performance with respect to the sample period. In light of Figure 8.4, the control cost of the co-design method is lower compared to high QoS for sample periods that are less than 400 ms. On the other hand, for $h_k \geq 400$, the co-design method performs almost the

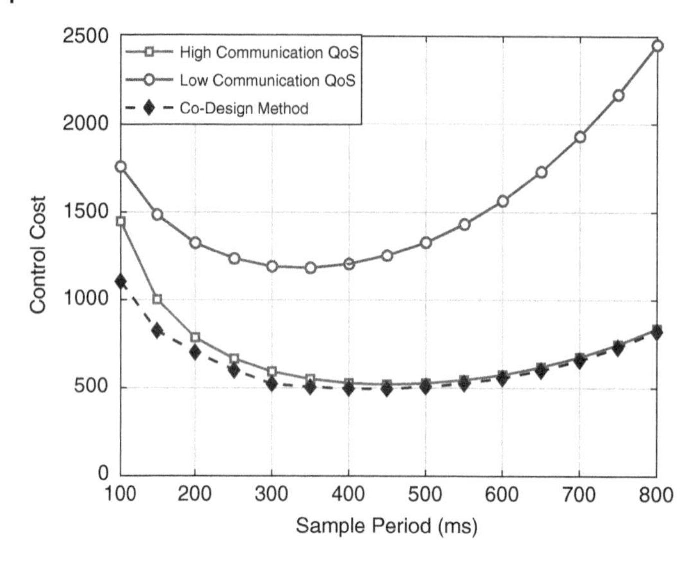

Figure 8.4 Control Cost Comparison.

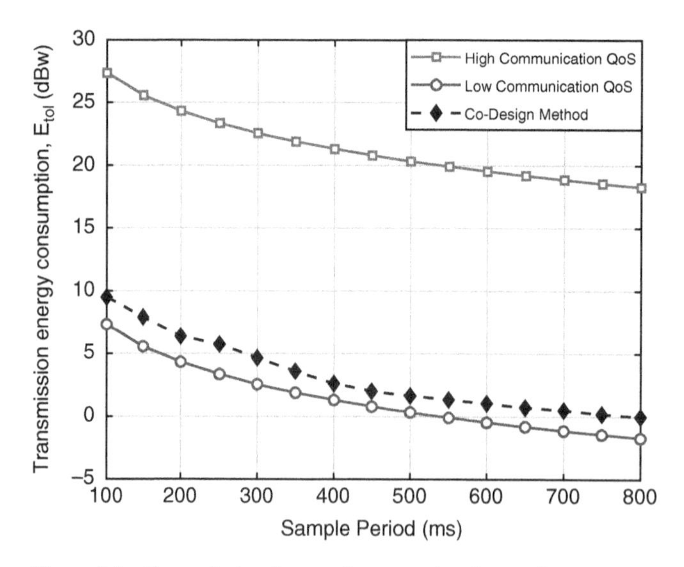

Figure 8.5 Transmission Energy Consumption Comparison.

same as high QoS in terms of the control cost. In summary, the co-design method performs very close to high QoS in terms of control cost. In addition, there is an optimal control cost, which is about 500 for sample period of 400 ms.

In Figure 8.5, the high QoS, low QoS, and co-design methods are compared in terms of transmission energy consumption performance with respect to the sample period. In light of Figure 8.5, the transmission energy consumption of co-design method only increases by about 5% compared to low QoS. On the other hand, the transmission energy consumption is reduced by about 80% compared to high QoS.

In summary, in light of Figure 8.4, and Figure 8.5, it can be concluded that the co-design method outperforms the conventional method in terms of transmission energy consumption and control cost. By using the co-design method, it becomes possible to maintain performance of real-time control as well as enabling efficient communication in terms of energy consumption.

8.6 Conclusions

Healthcare applications are important application areas of wireless communication and control systems. The first healthcare applications start with sensing and monitoring basic physiological information of patients. With the development of wireless communications and control, new application areas appeared in the concept of healthcare. Some promising applications such as remote surgery and remote diagnosis have been realized in the concept of both communication and control systems. In this chapter, a brief history of healthcare applications has been discussed. A general system diagram has been provided and explained in detail for real-time wireless control in healthcare. Technical requirements are explained to realize real-time communication and control over wireless links. Design aspects of wireless communication and control systems have been discussed by giving primary emphasis to communication-control co-design approaches to achieve relatively better overall system performance. Lastly, the system model for co-design approach is explained and simulation results have been discussed to highlight the benefits of co-design in terms of communication and control performance.

References

Aijaz, A., Dohler, M., Aghvami, A. H., Friderikos, V., and Frodigh, M. (2016). Realizing the tactile internet: haptic communications over next generation 5g cellular networks, *IEEE Wireless Communications*, 24(2), p. 82–89.

Akyildiz, I. F., Melodia, T., and Chowdhury, K. R. (2007). A survey on wireless multimedia sensor networks, *Computer networks*, 51(4), p. 921–960.

Akyildiz, I. F., Su, W., Sankarasubramaniam, Y., and Cayirci, E. (2002). Wireless sensor networks: a survey, *Computer networks*, 38(4), p. 393–422.

Akyildiz, I. F. and Vuran, M. C. (2010). *Wireless sensor networks*, volume 4. Hoboken, NJ: John Wiley & Sons.

Antonakoglou, K., Xu, X., Steinbach, E., Mahmoodi, T., and Dohler, M. (2018). Toward haptic communications over the 5g tactile internet, *IEEE Communications Surveys & Tutorials*, 20(4), p. 3034–3059.

Araújo, J., Mazo, M., Anta, A., Tabuada, P., and Johansson, K. H. (2014). System architectures, protocols and algorithms for aperiodic wireless control systems, *IEEE Transactions on Industrial Informatics*, 10(1), p. 175–184.

Bello, L. L., Åkerberg, J., Gidlund, M., and Uhlemann, E. (2017). Guest editorial special section on new perspectives on wireless communications in automation: From industrial monitoring and control to cyber-physical systems, *IEEE Transactions on Industrial Informatics*, 13(3), p. 1393–1397.

Boubaker, O. and Iriarte, R. (2017). The inverted pendulum in control theory and robotics: from theory to new innovations, volume 111. IET.

Cai, S. and Lau, V. K. (2018). Zero mac latency sensor networking for cyber-physical systems, *IEEE Transactions on Signal Processing*, 66(14), p. 3814–3823.

Chang, B., Zhao, G., Zhang, L., Imran, M. A., Chen, Z., and Li, L. (2019). Dynamic communication qos design for real-time wireless control systems, *IEEE Sensors Journal*.

ChinaDaily (2016). China performs first 5g-based remote surgery on human brain, March 18, http://www.chinadaily.com.cn/a/201903/18/WS5c8f0528a3106c65c34ef2b6.html.

Cizmeci, B., Xu, X., Chaudhari, R., Bachhuber, C., Alt, N., and Steinbach, E. (2017). A multiplexing scheme for multimodal teleoperation, *ACM Transactions on Multimedia Computing, Communications, and Applications (TOMM)*, 13(2), p. 21.

Durisi, G., Koch, T., and Popovski, P. (2016). Toward massive, ultrareliable, and low-latency wireless communication with short packets, *Proceedings of the IEEE*, 104(9), p. 1711–1726.

Eid, M., Cha, J., and El Saddik, A. (2010). Admux: An adaptive multiplexer for haptic–audio–visual data communication, *IEEE Transactions on Instrumentation and Measurement*, 60(1), p. 21–31.

El Kalam, A. A., Ferreira, A., and Kratz, F. (2015). Bilateral teleoperation system using qos and secure communication networks for telemedicine applications, *IEEE Systems Journal*, 10(2), p. 709–720.

Hachisu, T. and Kajimoto, H. (2016). Vibration feedback latency affects material perception during rod tapping interactions, *IEEE transactions on haptics*, 10(2), p. 288–295.

Marescaux, J., Leroy, J., Rubino, F., Smith, M., Vix, M., Simone, M., and Mutter, D. (2002). Transcontinental robot-assisted remote telesurgery: feasibility and potential applications, *Annals of surgery*, 235(4), p. 487.

Miao, Y., Jiang, Y., Peng, L., Hossain, M. S., and Muhammad, G. (2018). Telesurgery robot based on 5g tactile internet, *Mobile Networks and Applications*, 23(6), p. 1645–1654.

Nilsson, J. (1998). Real-Time Control Systems with Delays. PhD thesis, Department of Automatic Control, Lund Institute of Technology (LTH).

Park, P., Araújo, J., and Johansson, K. H. (2011). Wireless networked control system co-design. In: *2011 International Conference on Networking, Sensing and Control*, p. 486–491. IEEE.

Park, P., Ergen, S. C., Fischione, C., Lu, C., and Johansson, K. H. (2017). Wireless network design for control systems: a survey, *IEEE Communications Surveys & Tutorials*, 20(2), p. 978–1013.

Park, P., Ergen, S. C., Fischione, C., Lu, C., and Johansson, K. H. (2018). Wireless network design for control systems: A survey, *IEEE Communications Surveys & Tutorials*, 20(2), p. 978–1013.

Patel, M. and Wang, J. (2010). Applications, challenges, and prospective in emerging body area networking technologies, *IEEE Wireless communications*, 17(1), p. 80–88.

Perez, M., Xu, S., Chauhan, S., Tanaka, A., Simpson, K., Abdul-Muhsin, H., and Smith, R. (2016). Impact of delay on telesurgical performance: study on the robotic simulator dv-trainer, *International journal of computer assisted radiology and surgery*, 11(4), p. 581–587.

Rasool, F. and Nguang, S. (2008). Networked control systems, *Lecture Notes in Control and Information Sciences*, 52(9), p. 318–323.

Schenato, L., Sinopoli, B., Franceschetti, M., Poolla, K., and Sastry, S. S. (2007). Foundations of control and estimation over lossy networks, *Proceedings of the IEEE*, 95(1), p. 163–187.

She, C., Yang, C., and Quek, T. Q. (2016). Cross-layer transmission design for tactile internet. In: *2016 IEEE Global Communications Conference (GLOBECOM)*, p. 1–6. IEEE.

Simsek, M., Aijaz, A., Dohler, M., Sachs, J., and Fettweis, G. (2016). The 5g-enabled tactile internet: applications, requirements, and architecture. In: *2016 IEEE Wireless Communications and Networking Conference*, p. 1–6. IEEE.

UTRA, G. (2006). Physical layer aspects for evolved universal terrestrial radio access (utra).

Yang, W., Durisi, G., Koch, T., and Polyanskiy, Y. (2014). Quasi-static multiple-antenna fading channels at finite blocklength, *IEEE Transactions on Information Theory*, 60(7), p. 4232–4265.

Zhang, Q., Liu, J., and Zhao, G. (2018). Towards 5g enabled tactile robotic telesurgery, *arXiv preprint arXiv:1803.03586*.

Zhao, G., Imran, M. A., Pang, Z., Chen, Z., and Li, L. (2018). Toward real-time control in future wireless networks: communication-control co-design, *IEEE Communications Magazine*, 57(2), p. 138–144.

9

Role of D2D Communications in Mobile Health Applications: Security Threats and Requirements

Muhammad Usman[1], Marwa Qaraqe[1], Muhammad Rizwan Asghar[2] and Imran Shafique Ansari[3]

[1]*Division of Information and Computing Technology, College of Science and Engineering, Hamad Bin Khalifa University (HBKU), Education City, 34110 Doha, Qatar*
[2]*Department of Computer Science, The University of Auckland, 1142 Auckland, New Zealand*
[3]*James Watt School of Engineering, University of Glasgow, G12 8QQ, UK*

Monitoring physiological signals in real time has become popular among patients in recent years. Mobile health applications provide a platform to connect sensors with smartphones and transmit the sensors' readings to web platforms designed for practitioners and patients. Due to the life-critical nature of health data, it becomes imperative to establish an extremely reliable and 24/7 available connection to web platforms. In this regard, device-to-device (D2D) communication has the ability to establish communication links in different indoor and outdoor scenarios, including when the patient's phone is out of the coverage of any mobile network or has partial coverage. D2D is an emerging paradigm that enables different mobile health applications for varying use cases and scenarios. This chapter highlights the security challenges related to D2D communication in the field of mobile healthcare platforms. In particular, the focus is to identify different phases of D2D communication and to highlight the security requirements and threats associated within each phase. In the end, we identify various solutions to different security threats.

9.1 Introduction

The technology has radically changed the way patients are treated in this modern era. According to a recent study (Enbysk 2015), the traditional ways of healthcare management are unable to handle the data of the rapidly growing population of the world. There is a need to identify smarter ways to handle this problem. The modern communication technologies and traditional healthcare processes can be integrated together in order to develop a better healthcare system. This includes wireless body area networks (WBANs), communication networks, data analytics, and blockchain technology. The sensors acquire health information such as heartbeat, blood pressure, blood sugar, or any deterioration in health, which can be potentially transmitted, using communication technologies, to remote servers accessible by healthcare professionals for monitoring, diagnosis, or treatment purposes. In

Engineering and Technology for Healthcare, First Edition.
Edited by Muhammad Ali Imran, Rami Ghannam and Qammer H. Abbasi.

this regard, the trend of utilising mobile applications to monitor ones' health is increasing day-by-day (Usman et al. 2018).

A mobile healthcare system monitors the aforementioned physiological signals in real time and transmits emergency notifications in case of crisis. In contrast, device-to-device (D2D) communication is one of the important transmission modes, which has the capability of connecting mobile health applications in in-coverage, out-of-coverage, and partial-coverage areas. The literature of D2D communication in mobile health applications primarily focuses on preserving the privacy of medical data and utilizing various encryption algorithms to decrease the chances of eavesdropping. However, the communication modes and the scenarios play an important role in health data transmission, which is in fact the core of mobile healthcare systems. For instance, in the case of partial coverage, the reliability of relay nodes critically impacts the reliability of the transmitted data. Hence, it is important to potentially understand the scenarios and use cases of D2D communication, security requirements, and threats associated with each scenario.

This chapter highlights the security challenges related to D2D communication in the field of mobile health. In particular, the focus is to identify different phases of D2D communication and highlight the security requirements and threats associated with each phase. Finally, we summarize some attempts in the literature to secure the D2D communication from the threats identified in Section 9.3.

9.2 D2D Scenarios for Mobile Health Applications

In the 3rd Generation Partnership Project (3GPP), D2D communication is standardised under the name of proximity services (ProSe), which presents different use cases and scenarios, based on the coverage area of cellular networks. These scenarios can be categorized into three distinct types, which are elaborated in Figure 9.1 (Raza et al. 2019).

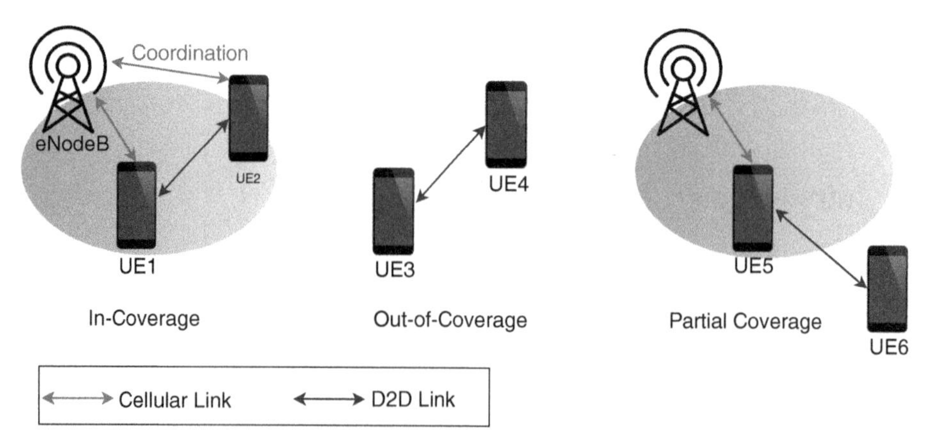

Figure 9.1 D2D communication use cases and scenarios based on the coverage area of the cellular network: "In-coverage" presents a scenario in which all UEs lie within the coverage area of the cellular network. "Out of coverage" signifies that all UEs lie outside the coverage area of the cellular network. "Partial coverage" presents a mixed scenario where some UEs lie within the coverage area while some lie outside the coverage area of the cellular network.

In-Coverage Scenario

In this scenario, all user devices could lie in the coverage area of the cellular network. The network operator is fully responsible for controling all functionalities of D2D communication, such as device discovery, device configuration, identity authentication, resource allocation, connection establishment and termination, access control, and security management. In this scenario, the D2D links are established in the same licensed spectrum on which the user equipments (UEs) are connected to the base station (BS). In other words, UE-to-UE communication and UE-to-BS communication share the same licensed spectrum.

Out-of-Coverage Scenario

In this scenario, all user devices are located outside the coverage area of the cellular network. The devices can establish D2D links without requiring any assistance from the infrastructure. This use case is the practical application of the disaster scenarios, where the cellular infrastructure is completely or partially damaged and a part of the network is offline. For instance, in Figure 9.1, UE3 and UE4 can establish D2D links and begin communication without requiring any support from the infrastructure. In this scenario, the UEs are responsible for controlling most of the functionalities of D2D communication, such as device discovery, device configuration, identity authentication, connection establishment and termination, access control, and security management. In 3GPP standardization procedures, communication in an out-of-coverage scenario is available only for public safety user devices, and commercial users cannot avail this service.

Partial-Coverage Scenario

In this scenario, some user devices are located outside the coverage area of the cellular network. The devices at the edge of the coverage area relay the information of out-of-coverage devices to the BS or core network. This helps to extend the coverage area of the network at the edge. For example, in Figure 9.1, UE5 acts as a relay node to extend the coverage of cellular networks to UE6. In this scenario, the network operator is fully responsible for controlling all functionalities of D2D communication, such as device discovery, device configuration, identity authentication, resource allocation, connection establishment and termination, access control, and security management, for both UE-to-UE and UE-to-BS communication.

9.3 D2D Security Requirements and Standardization

This section details the security threats and requirements for different D2D scenarios discussed in Section 9.2.

9.3.1 Security Issues on Configuration

9.3.1.1 Configuration of the ProSe Enabled UE

In order to perform device discovery, the ProSe-enabled UEs first need to be configured. The configuration data includes proximity criteria and radio resource configuration. This configuration must not be prone to manipulation by anybody, other than the authorized source, in order to prevent possible radio interference. The authorized sources include Host Public Land Mobile Network (HPLMN) and Visiting PLMN (VPLMN) operators. Any other source is not authorised to configure ProSe-enabled UE.

Security Threats

An attacker can maliciously configure the UE with false configuration data, which can deplete the radio resources. Moreover, an attacker can maliciously delete the UE data, stopping the UE from performing an operation. In addition, an attacker may eavesdrop the configuration data and redistribute it to unauthorised parties.

Security Requirements

The transmission of configuration data between the UE and the authorized configuration server must be mutually authenticated and protected for integrity, confidentiality, and replay. The configuration data must be stored in the UE securely.

9.3.2 Security Issues on Device Discovery

There are two possible methods to perform device discovery. One is in the out-of-coverage scenario, when UEs get no assistance from the network, and the other is in the partial-coverage and in-coverage scenarios, where infrastructure assists the UEs in device discovery. In scenarios when the network does not assist device discovery, UEs transmit direct request and response messages to each other to perform device discovery. While in the scenarios of network interaction, the messages are transmitted via a network entity, such as ProSe Function.

9.3.2.1 Direct Request and Response Discovery

This is the scenario of direct discovery, where the UEs can discover each other without any support from the infrastructure. A ProSe-enabled UE broadcasts its identity, which can be received by other ProSe-enabled UEs, which analyse the received identity and decide if any UEs of discoverer's interest are in their proximity.

Security Threats. An attacker, pretending to be a ProSe-enabled UE, may maliciously transmit discovery request and response messages to other UEs, which are not even entitled to perform D2D communication. This may lead to resource scarceness in ProSe-enabled UEs. In addition, an attacker may modify the discovery request and response messages, which may lead to incorrect discovery. Moreover, an attacker may replay the discovery request and response messages, which may lead to unauthorized use of ProSe discovery services. Furthermore, an attacker may obtain the International Mobile Subscriber Identity (IMSI) of a UE by intercepting the data packets of discovery request messages. Currently, there is no mechanism to guarantee that a UE that is discovering other UEs is in fact in the proximity of other UEs. So, an attacker who listens to the discovery information can broadcast it somewhere else. Consequently, a UE can falsely believe that other UEs are within the proximity, where actually there are none.

Security Requirements. The device request and discovery messages must be protected for preserving integrity and defending against replay. The UE that receives these messages should verify the authenticity of the source. Spatial replay protection of discovery messages must be guaranteed. The transmission of UE identity (IMSI) must be protected for preserving both integrity and confidentiality.

9.3.2.2 Open Direct Discovery

There are two types of direct discovery, open and restricted. In open direct discovery, a UE can be discovered by the other UEs, without any authorization from the UE being discovered. It means that the identity of the UE being discovered is known to all UEs. In restricted direct discovery, there is a requirement of explicit permission from the UE being discovered. It means that the UE must be authorized to discover a particular UE.

Security Threats. In the absence of any protection for the open direct discovery, there is a possibility of replay attack by a rogue UE by receiving the discovery information broadcast from the ProSe-enabled UEs. In this way, an attacker can easily impersonate other UEs and replay the discovery information, so that the discoverable UEs can be found when they actually are not in proximity.

Security Requirements. There should be measures in place to detect the replay and impersonation attacks in open direct discovery.

9.3.2.3 Restricted Direct Discovery

With restricted discovery, a ProSe-enabled UE cannot be discovered by other UEs without explicit permission from the UE being discovered. In addition, a UE needs to be authorized to discover other UEs.

Security Threats A UE that is authorised to discover others can easily track the broadcasting UE based on its broadcast identity and broadcast message contents over time. Similar to open discovery, impersonation and replay attacks are the other possible threats to restricted direct discovery.

Security Requirements The identities that are broadcast over the air interface must not be understood by the currently unauthorized UEs. Tracking of a UE, based on its broadcast contents over time, must be prevented. In addition, impersonation and replay attacks must be prevented. The restricted discovery messages must be protected for integrity and confidentiality while keeping the processing load reasonable for the receiving UE.

9.3.2.4 Registration in Network-Based ProSe Discovery

In order to begin D2D communication, each ProSe-enabled UE must get a valid ProSe ID from the operator. In the network-assisted discovery, a ProSe-enabled UE, willing to start a proximity service, transmits a registration request to the ProSe server to obtain the ProSe ID. The ProSe server validates the subscription of the UE for the requested service and assigns a ProSe ID for the application instance.

Security Threats. There might be a denial-of-service (DoS) attack in attempting to maliciously register an unauthorized UE. An attacker transmits a registration request to the ProSe server via eNodeB (Evolved Node B) and MME (mobility management entity). The eNodeB and MME, having no right to check the authenticity of the requesting UE, forwards this request to the ProSe server. The ProSe server, after performing a subscription test, refuses to allocate a ProSe ID. The attacker continually transmits the registration requests, and ProSe server refuses every request. As a result, this process brings the DoS attack and degrades the performance of the ProSe server and the network.

Security Requirements. The network must take possible measures to detect the DoS attack and halt it.

9.3.3 Security Issues on One-to-Many Communications

In one-to-many communication, a UE can communicate with a large group of UEs in the proximity. The key issues for performing a secure one-to-many communication are further classified in the following paragraphs.

9.3.3.1 One-to-many communications between UEs

One-to-many communication is an obligatory requirement for ProSe UEs. The UEs must be able to begin D2D communication without discovering the receiving UEs. As a result, encrypted data packets of one-to-many communication must be successfully decrypted by other group members without knowing in advance to which group members these packets are intended for.

Security Threats. An eavesdropper may eavesdrop the data packets exchanged between the UEs and obtain the actual content. Moreover, an eavesdropper may modify the data packets exchanged between the UEs without any detection by the transmitting and the receiving UEs. There is an increased chance of replay attacks in one-to-many communication. In this regard, a UE may have some problem to receive particular information, then later this information might be replayed while the receiving UE may consider it as a fresh transmission. The integrity threats in one-to-many communication seem to be similar to the eNodeB-to-UE communication in conventional cellular networks.

Security Requirements. The data exchanged between the UEs must be protected without requiring interaction between them. Moreover, the confidentiality must be protected for in-coverage, out-of-coverage, and partial coverage scenarios. In addition, the security of one-to-many communication must scale with the group size.

9.3.3.2 Key Distribution for Group Communications

The one-to-many communication must be secured by some key generation and distribution mechanism. This requires a secure procedure to share the key among the members of a ProSe group. There is a possibility of network supported key distribution among group members, which can be utilised when UEs are in or out of coverage of the network. In addition, for out-of-coverage scenario, there must be an extra level of protection. In the group owner mode, the group owner must generate and distribute a session key to its group members. This session key limits the exposure of the pre-shared key, which was shared when ProSe enabled UE was in the network coverage. In decentralised mode, the session key must be generated and distributed by the group member who is initiating the communication.

Security Threats. An attacker can eavesdrop the shared key and modify all the communication protected with that key. Moreover, the attacker can modify the key itself, which can lead to the following security threats.

- The attacker can share the modified key with some group member. This leads to DoS attack and UE can no longer communicate with each other.
- The ProSe group members can share the modified key and the attacker can monitor and modify all communication among ProSe group members.

An attacker, having exposure of the shared key, can perform replay attacks. In case the same shared key is utilised for longer periods of time, the attacker may deduce the security information from the data packets and obtain the actual content.

Security Requirements Both the shared and the session keys must be protected for ensuring both integrity and confidentiality, especially when UEs are out of the network coverage. There must be some secure mechanism to authenticate the network entity distributing the shared key and the group member/owner distributing the session key. The shared keys of past and future sessions must be securely stored in the UEs. In addition, the distribution of session key must support the late entry in the group.

9.3.4 Security Issues on One-to-One Communication

9.3.4.1 One-to-One ProSe Direct Communication

The one-to-one security must be comparable with the one provided by the existing 3GPP network. In existing 3GPP system, each connected UE uses a separate security context. The same must be applied to ProSe communication.

Security Threats

If same security context is utilised then a ProSe enabled UE can decrypt the communication between two other ProSe enabled UEs.

Security Requirements

A ProSe enabled UE must utilise different security contexts for one-to-one communication with different UEs.

9.3.4.2 One-to-One ProSe Direct Communication

The direct communication path between two ProSe UEs can be established with or without the support of UE-Network relay or UE-UE relay. The UEs need to be authenticated to establish a secure communication path, when one or both UEs are out of network coverage.

Security Threats. If the relay node generates a key for securing communication path between UEs, a misbehaving relay can eavesdrop or modify the communication data. If session keys are not updated for a long time, the attacker can try to recover the communication data.

Security Requirements. The UEs must be mutually authenticated. In addition, the communication between UEs must offer both integrity and confidentiality. Moreover, the keys must be frequently updated to avoid key stream reuse.

9.3.5 Security Issues on ProSe Relays

A ProSe-enabled UE can act as a relay node to establish a communication path between two UEs (UE-UE relay) or between the network and a UE (UE-network relay) if the UE is out of the network coverage.

9.3.5.1 Maintaining 3GPP Communication Security through Relay

The level of security provided in the UE-network or the UE-UE relay must be comparable with the one provided in the 3GPP system. The relay must not be able access the contents it is relaying.

Security Threats. A misbehaving relay can impersonate the communication data of a relayed user, eavesdrop on the communication data of a relayed user, replay the user/network communication, prevent data communication between UEs, and undermine the privacy of the user.

Security Requirements. The attacker must be prevented from impersonating and eavesdropping on the user's data, and from modifying and replaying communication. In addition, there must be an authentication mechanism to ensure that the relay is offering a legitimate service.

9.3.5.2 UE-Network Relay

According to the 3GPP standard, the UE-Network relay must support both unicast and Evolved Multimedia Broadcast Multicast Services (eMBMS) relay for remote UEs that are not in the network coverage.

Security Threats. A remote UE can connect to a fraudulent UE-Network relay, or a UE-Network relay can provide the relaying services to a fraudulent remote UE. Moreover, the remote UE and UE-Network relay exchange discovery requests and response messages for device discovery. As a result, there can be MitM (Man in the Middle) attacks, where an attacker can modify the discovery or data messages between the remote UE and the UE-Network relay.

Security Requirements. Both the UE-Network relay and remote UE must be authorized and be legitimate. In addition, the discovery request and response messages must be protected. There must be a secure mechanism to mutually authenticate the remote UE and UE-Network relay.

9.3.5.3 UE-to-UE Relay

Two ProSe-enabled UEs can establish a one-to-one communication path among them via a relay node. For this, the UE-to-UE relay and remote UE need to first discover each other by exchanging discover requests and response messages.

Security Threats. A remote UE can connect to a fraudulent UE-to-UE relay or a UE-to-UE relay, can provide the relaying services to a fraudulent remote UE. Or a remote UE1 can connect to a fraudulent UE2 via UE-to-UE relay. This may lead to modifying and replaying the UE-to-UE discovery signalling.

Security Requirements. The attacker must be prevented from making replay and impersonation attacks on discovery messages. In addition, the connection of UE1 via UE-to-UE relay to fraudulent UE2 must be prevented. Moreover, both UE-to-UE relay and remote UEs must be mutually authenticated and authorized.

9.4 Existing Solutions

In this section, we discuss the individual solutions presented in the literature to secure D2D communications and categorize them in the following broad areas.

9.4.1 Key Management

Zhang, Chen, Hu, and Qian (2016), introduce SeDS, which is a secure data sharing protocol. Using SeDS, they address the problem of secure data exchange in D2D communications.

Specifically, they assume the presence of a trusted authority to sign public keys of nodes. Using public keys, a secure communication channel is established for guaranteeing confidentiality and integrity. More technically, communicating nodes exchange their public keys to establish a symmetric key. This key can ensure confidentiality and integrity but cannot provide non-repudiation of the receiver. To this end, Zhang, Hu, and Zhang require that the receiver transmit a key request message in order to retrieve a key that is utilized to encrypt the requested data. This key request message can ensure non-repudiation of the receiver. There are two major issues with this approach. First, it requires the presence of a trusted third party. Second, revoking a node's public key is not trivial, and it remains unclear how key revocation can be integrated to exclude the possibility of compromised nodes.

Shen et al. (2014) investigate security requirements for D2D communications and present a key exchange protocol that enables two nodes to establish a shared secret without prior knowledge. They also integrate their protocol into the Wi-Fi direct protocol and provide an implementation for Android smartphones. However, the main issue with this scheme is a strong assumption of mutual authentication between nodes through out-of-band means. In other words, this approach can work well for scenarios where users can manually authenticate each other using some secret message as long as the secret message remains protected from adversaries and all the nodes do not leak any private information.

Goratti et al. (2014) propose a protocol for securing D2D communication. The authors targeted Public Safety (PS) users with out-of-coverage UEs. The communication between UEs (D2D users) is performed in licensed bands. The team leaders of PS groups, regarded as b-UEs, broadcast beacon messages to start D2D communication. To embed security features, the authors propose a key exchange scheme derived from sensor networks. The keys are randomly pre-distributed among UEs and indexed in subfield of beacon messages.

Alam et al. (2014) propose security architecture for D2D communication in LTE-A networks along with security threats and requirements. In their work, the authors consider three use cases and scenarios for D2D communication. Based upon these scenarios, they propose three solutions to secure D2D communication by reusing the existing security mechanisms. They propose network-controlled key distribution in traffic offload scenarios, where UEs lie in the coverage area of the network with no user application for D2D communication (D2D is seamless to users and controlled by the network). For the second scenario, where UEs lie within the coverage area of the network and loaded with user application, the authors propose network-and-application-controlled authentication. For the third scenario, where UEs lie outside the coverage area of the network, a pre-shared-key based security scheme is proposed.

Sun et al. (2014) present SYNERGY, a game-theoretic approach for a cooperative key generation in wireless networks, which could be used for securing D2D communication. The main idea behind SYNERGY is to partition all the nodes into multiple disjoint coalitions. There is an assumption that a node will help the nodes in its coalition to establish secret keys and to receive help in return. They assume key generation that is based on physical characteristics of the channel. The major issue with this approach is the dynamic nature of mobile nodes in D2D communication, which effectively limits the applicability of key establishment based on physical layer characteristics. In other words, key establishment based on physical layer characteristics can work when communication nodes are stationary or at least not moving significantly.

Xi et al. (2014) have stated that generating keys individually on different communication nodes without any exchange is desirable, though challenging. Some recent works have proposed extracting keys from the measurements of physical layer random variations of a wireless channel. But these existing Channel State Information (CSI)-based key extraction methods usually utilize the measurements of individual subcarriers (Xi et al. 2014). However, Xi et al. (2014) mention the real world experimental results, which demonstrate that the CSI measurements from near-by subcarriers have strong correlations and hence a generated key may have a large proportion of repeated bit segments. Therefore, the attackers can crack the key faster, thereby reducing the security level of the generated keys. To overcome this issue, Xi et al. (2014) have proposed a fast secret key extraction protocol that utilizes a validation recombination mechanism to obtain consistent secret keys from CSI measurements of all subcarriers. The authors implemented this protocol using off-the-shelf 802.11n devices and evaluate its performance based on experiments demonstrating this protocol to be safer and more effective than the state-of-the-art approaches.

9.4.2 Routing

Panaousis et al. (2014) propose the secure message delivery (SMD) protocol to securely transport messages from source to destination in a multi-hop D2D network. Apart from security, the authors consider energy consumption and quality of service (QoS) as a part of SMD. The authors utilize application-level techniques for malware detection such as anti-virus. Every node in the route inspects the delivered messages for potential malware. Based on this inspection, a path with more inspection capabilities is chosen along with some emphasis on energy consumption and QoS.

Jung, Festijo, and Peradilla (2014) propose a protocol for securing D2D communications. The idea is to utilize a group key agreement method that could be combined with the routing protocol. If needed, the group key could be updated periodically when a new node joins the network. To perform authentication, the protocol assumes a node certificate issued by a certification authority. Like other approaches (such as Zhang et al. (2016)), two major issues include the presence of trusted certificate authority and lack of support for certificate revocation.

9.4.3 Social Trust and Social Ties

Chen et al. (2013, 2015) propose social trust and social reciprocity based D2D communication, which exploits social ties in human social networks to enhance security and cooperation in D2D UEs. The authors propose coalition-based game solution to formulate a relay selection algorithm with support from a cellular network. The relay members are selected based upon their social behavior towards other members.

Ometov et al. (2016) propose a social awareness layer in D2D communication to build trust among D2D UEs. Before establishing D2D clusters, the UEs examine the social behavior patterns and interpersonal relationship of humans, thus forming trusted user groups. However, the authors claim that this trust and social-aware cooperation between UEs and with the network operators remains conditional to the incentives provided to participating UEs by the network operators. The authors identify three kinds of possible user incentives that apply to different D2D scenarios; Pragmatic incentives (throughput gain, energy efficiency, latency gain), Indirect incentives (economic incentives), and Social incentives (lend resources to friends and family).

Wang and Wu (2015), Wang et al. (2015) leverage social relationship among UEs to increase the secrecy rate of network-controlled D2D communication. The authors propose the selection of friendly jamming partners to minimize the social outcasts. The jamming partner is selected based upon its social trust. A heuristic algorithm is presented for allocating transmit-power to both source and the jammer. The authors assess the impact of social trust of cooperative nodes for a different number of jamming partners.

Wang, Wu, and Stuber (2016) consider the problem of reliability and secrecy enhancement for wireless content being shared between two communicating nodes. By exploiting social characteristics of multiple nodes in the presence of multiple independent eavesdroppers, they first investigate the impact of mobility for source node selection for transmission reliability, within a cooperative wireless network. Furthermore, they address social-tie-based jammer node selections for cooperative jamming to provide secrecy, while allocating power appropriately to the source node and the cooperative jammer node to maximize the worst-case ergodic secrecy rate. Specifically, in Wang, Wu, and Stuber (2016), the source transmitter is selected by first narrowing the potential link set to provide desired transmission success rate while considering the physical channel and social characteristics jointly. Moreover, instead of focusing on full CSI, they investigated more practical case where only statistical CSI was known. Their optimisation is based on the upper and lower bounds to the secrecy rate thereby simplifying the power optimisation problem without noticeable performance loss.

Ometov et al. (2016) propose information security protocols to secure information exchange between D2D UEs and associated wearable devices. They first utilise game-theoretic mechanisms to form user clusters based on both spatial and social proximity. The information security protocols are then employed to secure information exchange among D2D UEs and associated wearable devices. In case of infrastructure presence, the group of D2D UEs can establish their own information security rules with conventional methods. However, in the absence of cellular infrastructure the proposed information security procedures facilitate secure data exchange and cluster joining and leaving.

The work in Wang, Cao, and Wu (2015) investigated methods to form clusters to overcome social issues in D2D communications. They form reliable cooperative clusters based on physical positions of candidate DUEs and further exploit their social relationships to improve secrecy rate. The cluster-assisted relay and jammer selection optimisation scheme in Wang, Cao, and Wu (2015) has two-fold objective of maximising the D2D User Equipment (DUE) secrecy rate under power constraint and satisfying the requisite Signal-to-Interference-plus-Noise Ratio (SINR) need of co-channel Cellular User Equipments (CUEs). To solve the non-convex optimisation problem, the authors proposed a Generalised Fractional Programming (GFP) method by utilising the Dinkelbach-type algorithm. They also analysed the impact of social trust and channel estimation error of these mobile nodes to demonstrate the robustness and reliability of the proposed approach. Numerical results demonstrate that the proposed scheme achieved better performance based on the underlying D2D cooperation.

The authors in Ometov et al. (2016) present a Proof of Concept (PoC) prototype implementation with information security protocols for socially aware D2D systems. The protocol suite enables secure data delivery for already communicating D2D UEs. In case of a connection with cellular systems, the D2D UEs can establish and manage their secure D2D network. However, in case of intermittent connectivity with the cellular systems, their proposed information security protocol allows a set of existing D2D UEs in the secure group to admit a previously unknown UE or exclude an existing UE from the group. The author's proposal is based on enabling the Trusted Social Association Framework (TSAF) to allow the formation of semi-supervised D2D-based proximate communities. In TSAF, all users have equal voting rights to include or exclude a UE.

9.4.4 Access Control

Yue et al. (2013) utilize interference of D2D communication as a tool against eavesdropping, where D2D communication causes interference to the eavesdropper. The D2D communication is introduced into the problem of information-theoretic secrecy of the cellular network. As the CSI of the eavesdropper is mostly unknown, the secrecy outage probability is considered as the secrecy requirement of the cellular communication. Based upon the outage probability, the authors find the optimal transmission power of the D2D communication pair.

9.4.5 Physical Layer Security

Jayasinghe et al. (2015) present a secure beam-forming design in order to prevent eavesdropping on Multiple-Input Multiple-Output (MIMO) D2D communication systems. Specifically, the devices communicate through a trusted relay that performs Physical-layer Network Coding (PNC) while multiple eavesdroppers try to intercept the device information. The beam-forming design is based on minimizing the mean square error of the D2D communications while employing SINR threshold constraints in order to prevent possible eavesdropping. The channel state information of the device-to-eavesdropper and relay-to-eavesdropper links are imperfect at the devices and relays, respectively. The channel estimation errors are assumed to be governed by the Gaussian Markov uncertainty model. Consequently, similar to the work in Wang, Cao, and Wu (2015), robust optimization problems were formulated considering the multiple access and broadcasting stages of the D2D communication systems. These problems came out to be non-convex; thereby two algorithms were proposed to solve them. The numerical analysis of Jayasinghe et al. (2015), the authors shared the convergence of their proposed algorithms' impact of the number of eavesdroppers on the performance, and the SINR distributions at eavesdroppers.

As shared above, D2D communication has been proposed to improve spectral efficiency. Zhang et al. (2014) consider physical-layer security in underlay D2D communications in presence of an eavesdropper. They state that benefiting from the underlay spectrum reuse, D2D users can contribute to the system secrecy capacity; on the other hand, D2D users may even interfere with the cellular users, thereby decreasing their secrecy capacity. Hence, they formulated this as a matching problem in the weighted bipartite graph and introduced the Kuhn-Munkres (KM) algorithm to provide an optimal solution. Simulation results demonstrated that the system secrecy capacity can be greatly improved by introducing D2D communications in underlying cellular networks.

In line with Zhang et al. (2014), the authors in Zhang, Cheng, and Yang (2016) investigate the cooperation issue based on spectrum sharing when employing the physical layer security concept in D2D communications underlying cellular networks. Initially, they derived the optimal joint power control solutions of the cellular communication links and D2D pairs in terms of the secrecy capacity under a simple cooperation case and further proposed a secrecy-based access control scheme via the best D2D pair selection mechanism. Then, they considered a more general case wherein multiple D2D pairs could access the same Resource Block (RB) and one D2D pair was permitted to access multiple RBs, thereby providing a novel cooperation mechanism for the investigated network. Furthermore, they formulated the proposed cooperation mechanism among cellular communication links and D2D pairs as a coalition-based game. Based on a newly defined Max-Coalition order in the constructed game, they proposed a merge-and-split-based coalition formation algorithm for cellular communication links and D2D pairs to achieve efficient and effective cooperation, leading to improved system secrecy rate and social welfare.

The same approach of optimizing the radio resource allocation to D2D and cellular UEs is considered in Zhang et al. (2016) aiming to maximize the secrecy capacity of D2D communications underlying Heterogeneous Networks (HetNets). The authors use Perron-Frobenius theory to derive a convex equivalent of the non-convex optimization problem of resource allocation. Then a heuristic algorithm is proposed based on the proximal theory. The authors compare the secrecy capacity performance under different QoS requirements.

Zhu et al. (2014) demonstrate that the D2D paradigm significantly improves security at the physical layer, by reducing information leakage to eavesdroppers from two relatively high-power transmissions to a single low-power hop. Henceforth, they derived the Secrecy Outage Probability (SOP) for D2D and cellular systems, and analysed the performance of D2D systems in presence of a multi-antenna eavesdropper. The cellular approach was only seen to have an advantage relatively in certain cases where the Access Point (AP) was equipped with a large number of antennas and perfect CSI.

D2D communication provides a promising technique for 5G wireless networks, supporting higher data rates (Ghanem and Ara 2015). Security of data transmissions over wireless clouds may place constraints on devices. Therefore, Ghanem and Ara (2015) targeted to provide an analytical framework for security at the physical layer and to define the constraints involved with cooperation in wireless clouds. In Ghanem and Ara (2015), two legitimate transmitters Alice and John cooperate to increase the reliable transmission rate received by their common legitimate receiver Bob in the presence of Eve (eavesdropper/illegitimate receiver). They proposed a distributed algorithm that allows the devices to opt whether to or not to cooperate and to adapt their optimal power allocation dependent on the cooperation framework selected. Moreover, they defined distance constraints to enforce the advantages of cooperation between devices in a wireless cloud.

Awan and Sezgin (2015) analyzed the issue of secure transmission over a caching D2D network. The model assumed in Awan and Sezgin (2015) allows end users to pre-fetch a part of popular contents in their local cache. Users request arbitrarily from the library of available files and interact with each other directly to share requested contents from the local cache, thereby jointly satisfying their demands. The transmission between the users needs to be secured as it is wiretapped by an external eavesdropper. For the assumed model, in Awan and Sezgin (2015) exploit the flexibility offered by local cache storage and establish a coding scheme that conforms to the demands of all users and delivers the contents securely. Relative to the insecure caching schemes, this coding scheme illustrated that for a large number of files and users, the loss incurred due to the imposed secrecy constraints is insignificant.

Robust secrecy rate optimization problems for a Multiple-Input Single-Output (MISO) secrecy channel with multiple D2D communications is investigated in Chu et al. (2015). Specifically, they assume that the D2D communication nodes share the same spectrum and assist in improving secrecy communications of the system by confusing eavesdroppers. On the other hand, the legitimate transmitter ensures that the D2D communicating nodes obtain their required rates. Moreover, the legitimate transmitter is assumed to have imperfect channel state information of different nodes. For such secrecy network, Chu et al. (2015) tackle two robust secrecy rate optimization problems: (a) robust power minimization problem, subject to the probability-based secrecy rate and the D2D transmission rate constraints; (b) robust secrecy rate maximization problem with the transmit power, the probabilistic-based secrecy rate, and the D2D transmission rate constraints. Owing to the non-convexity of robust beam-forming design based on two statistical channel uncertainty models, Chu et al. (2015) proposed two conservative approximation approaches based on Bernstein-type inequality and S-Procedure to address these robust optimization problems. Simulation results were provided to validate the performance of these two conservative approximation methods, where it is demonstrated that the Bernstein-type inequality based approach outperforms the S-Procedure approach in terms of achievable secrecy rates.

Liu et al. investigate secure D2D communication in energy harvesting large-scale cognitive cellular networks (2016). Specifically, the energy-constrained D2D transmitter harvests energy from multi-antenna-equipped Power Beacons (PBs), and communicates with the corresponding receiver utilizing the spectrum of the primary Base Stations (BSs). They introduced a power transfer model and an information signal model to enable wireless energy harvesting and secure information transmission. In the power transfer model, three Wireless Power Transfer (WPT) policies are proposed: 1) Co-operative Power Beacons (CPB) power transfer, 2) Best Power Beacon (BPB) power transfer, and 3) Nearest Power Beacon (NPB) power transfer. To characterize the power transfer reliability of the three proposed policies, the authors derived new expressions for the exact power outage probability. In the information signal model, they presented a new comparative framework with two receiver selection schemes: 1) Best Receiver Selection (BRS), where the receiver with the strongest channel is selected; and 2) Nearest Receiver Selection (NRS), where the nearest receiver is selected. To assess secrecy performance, they derived new analytical expressions for the secrecy outage probability and the secrecy throughput, considering the two receiver selection schemes utilizing the proposed WPT policies. It was demonstrated that: 1) secrecy performance improves with increasing densities of PBs and D2D receivers due to larger multiuser diversity gain; 2) CPB achieves better secrecy performance than BPB and NPB but consumes more power; and 3) BRS achieves better secrecy performance than NRS but demands more instantaneous feedback and overhead (Liu et al. 2016).

D2D communication underlying cellular network is a promising technology to improve network resource utilization (Ma et al. 2015). In D2D-enabled cellular networks, interference generated by D2D communications is usually viewed as an obstacle to cellular communications. However, Ma et al. (2015) presented a new perspective on the role of D2D interference by taking security issues into consideration. They considered a large-scale D2D-enabled cellular network with eavesdroppers overhearing cellular communications. Using stochastic geometry, they modeled such a network and analyzed the SINR distributions, connection probabilities, and secrecy probabilities of both the cellular and D2D links. They proposed two criteria for guaranteeing performance of secure cellular communications, namely the strong performance guarantee criteria and weak performance guarantee criteria. Based on the obtained analytical results of link characteristics, Ma et al. (2015) designed optimal D2D link-scheduling schemes under these two criteria respectively. Both analytical and numerical results in Ma et al. (2015) proved that the interference from D2D communications can improve physical layer security of cellular communications while simultaneously creating extra transmission opportunities for D2D users.

9.4.6 Network Coding

Pahlevani et al. (2014) propose network coding to secure network-controlled D2D communications. Instead of using a store-and-forward paradigm, the authors introduce a compute-and-forward paradigm in their work to enhance reliability and security. Their network-coding-based proposal includes the transmission of a file in parts, where a part is transmitted by the cellular network and other parts are transmitted by direct D2D communications. This makes it difficult for an eavesdropper to decode the information. The authors further propose the encryption of the coding coefficients, in case the attacker overhears all packets from cellular and D2D networks.

9.5 Conclusion

This chapter has divided D2D communications for mobile health application into different phases. For each phase, associated security threats and requirements have been identified. Further, we have provided a thorough literature review, existing solutions have been critically analyzed and investigated in order to fulfil the security requirements. We believe that a secure D2D network is pivotal in realising a reliable, secure, and 24/7-available mobile healthcare platform.

References

Alam, M., D. Yang, J. Rodriguez, and R.A. Abd-alhameed, (2014). Secure device-to-device communication in LTE-A, *IEEE Communications Magazine*, 52(4), p. 66–73.

Awan, Z. H., and A. Sezgin, A. (2015). Fundamental limits of caching in D2D networks with secure delivery, in 2015 IEEE International Conference on Communication Workshop (ICCW), p. 464–469.

Chen, X., B. Proulx, X. Gong, and J. Zhang. (2013). Social trust and social reciprocity based cooperative D2D communications, in Proceedings of the Fourteenth ACM International Symposium on Mobile Ad Hoc Networking and Computing. MobiHoc '13, ACM. New York, NY, USA, p. 187–196.

Chen, X., B. Proulx, X. Gong, and J. Zhang. (2015). Exploiting social ties for cooperative D2D communications: a mobile social networking case, *IEEE/ACM Transactions on Networking*, 23(5), p. 1471–1484.

Chu, Z., K. Cumanan, M. Xu, and Z. Ding. (2015). Robust secrecy rate optimisations for multiuser multiple-input-single-output channel with device-to-device communications, *IET Communications*, 9(3), 396–403.

Enbysk, L. (2015). How smart technology is improving public health, http://smartcitiescouncil.com/article/how-smart-technology-improving-public-health. (Accessed 15 November 2019.)

Ghanem, S.A.M., and M. Ara. (2015). Secure communications with D2D cooperation, in 2015 International Conference on Communications, Signal Processing, and their Applications (ICCSPA), p. 1–6.

Goratti, L., G. Steri, K.M. Gomez, K. M., and G. Baldini. (2014). Connectivity and security in a D2D communication protocol for public safety applications, in 2014 11th International Symposium on Wireless Communications Systems (ISWCS), p. 548–552.

Jayasinghe, K., P. Jayasinghe, N. Rajatheva, and M. Latva-aho (2015). Physical layer security for relay assisted MIMO D2D communication, in 2015 IEEE International Conference on Communication Workshop (ICCW), p. 651–656.

Jung, Y., E. Festijo, and M. Peradilla. (2014). Joint operation of routing control and group key management for 5G ad hoc D2D networks, in 2014 International Conference on Privacy and Security in Mobile Systems (PRISMS), p. 1–8.

Liu, Y., L. Wang, S.A.R. Zaidi,M. Elkashlan, and T.Q. Duong. (2016). Secure D2D communication in large-scale cognitive cellular networks: a wireless power transfer model, *IEEE Transactions on Communications*, 64(1), 329–342.

Ma, C., J. Liu, X. Tian, H. Yu, Y. Cui, and X. Wang. (2015). Interference exploitation in D2D-enabled cellular networks: a secrecy perspective, *IEEE Transactions on Communications*, 63(1), p. 229–242.

Ometov, A., E. Olshannikova, P. Masek, T. Olsson, J. Hosek, S. Andreev, and Y. Koucheryavy. (2016). Dynamic trust associations over socially-aware D2D technology: a practical implementation perspective, *IEEE Access*, 4, p. 7692–7702.

Ometov, A., A. Orsino, L. Militano, G. Araniti, D. Moltchanov, and S. Andreev. (2016). A novel security-centric framework for D2D connectivity based on spatial and social proximity, *Computer Networks*, 107(pt.2), p. 327–338.

Ometov, A., A. Orsino, L. Militano, D. Moltchanov, G. Araniti, E. Olshannikova, G. Fodor, et al. (2016). Towards trusted, social-aware D2D connectivity: bridging across technology and sociality realms, *IEEE Wireless Communications*, 23(4), p. 103–111.

Pahlevani, P., M. Hundebøll, M.V. Pedersen, D. Lucani, H. Charaf, F.H.P. Fitzek, H. Bagheri, and M. Katz. (2014). Novel concepts for device-to-device communication using network coding, *IEEE Communications Magazine*, 52(4), p. 32–39.

Panaousis, E., T. Alpcan, H. Fereidooni, and M. Conti. (2014). *Secure message delivery games for device-to-device communications*, Springer International Publishing, p. 195–215.

Raza, U., M. Usman, M.R. Asghar, I.S. Ansari, and F. Granelli. (2019). Integrating public safety networks to 5G: applications and standards, in In: *Enabling 5G communication systems to support vertical industries*. Edited by Imran, M.A., Sambo, Y.A., and Abbas, Q.H. Hoboken, NJ: John Wiley & Sons, 233–251.

Shen, W., W. Hong, X. Cao, B. Yin, D.M. Shila, and Y. Cheng. (2014). Secure key establishment for device-to-device communications, *IEEE Globecom* abs/1410.2620.

Sun, J., X. Chen, J. Zhang, Y. Zhang, and J. Zhang. (2014). SYNERGY: A game-theoretical approach for cooperative key generation in wireless networks in IEEE INFOCOM 2014 - IEEE Conference on Computer Communications, p. 997–1005.

Usman, M.R. Asghar, and F. Granelli. (2018). 5G and D2D communications at the service of smart cities, *Transportation and power grid in smart cities: communication networks and services* p. 147–169. Wiley Telecom. https://ieeexplore.ieee.org/document/8654114.

Wang, L., C. Cao, and H. Wu. (2015). Secure inter-cluster communications with cooperative jamming against social outcasts, *Computer Communications*, 63, p. 1–10.

Wang, L., and H. Wu. (2015). Jamming partner selection for maximising the worst D2D secrecy rate based on social trust, *Transactions on Emerging Telecommunications Technologies*.

Wang, L., H. Wu, L. Liu, M. Song, and Y. Cheng. (2015). Secrecy-oriented partner selection based on social trust in device-to-device communications. in 2015 IEEE International Conference on Communications (ICC), p. 7275–7279.

Wang, L., H. Wu, and G.L. Stuber. (2016). Cooperative jamming aided secrecy enhancement in P2P communications with social interaction constraints, *IEEE Transactions on Vehicular Technology* **PP**(99), 1–1.

Xi, W., X. Li, C., Qian, J. Han, S. Tang, J. Zhao, and K. Zhao. (2014). KEEP: fast secret key extraction protocol for D2D communication, in IEEE 22nd International Symposium of Quality of Service, IWQoS 2014. Hong Kong, China, May 26-27, 2014, p. 350–359.

Yue, J., C. Ma, H. Yu, an W. Zhou. (2013). 'Secrecy-based access control for device-to-device communication underlaying cellular networks, *IEEE Communications Letters*, 17(11), p. 2068–2071.

Zhang, A., J. Chen, R.Q. Hu, and Y. Qian. (2016). SeDS: Secure data sharing strategy for D2D communication in LTE-advanced networks, *IEEE Transactions on Vehicular Technology*, 65(4), p. 2659–2672.

Zhang, H., T. Wang, T., L. Song, Z. Han. (2014). Radio resource allocation for physical-layer security in D2D underlay communications. In: 2014 IEEE International Conference on Communications (ICC), p. 2319–2324.

Zhang, K., M. Peng, P. Zhang, and X. Li. (2016). Secrecy-optimized resource allocation for device-to-device communication underlaying heterogeneous networks. *IEEE Transactions on Vehicular Technology*, PP(99), p. 1–1. 66 (2) Feb. 2017). https://ieeexplore.ieee.org/document/7467574.

Zhang, R., X. Cheng, and L. Yang. (2016). Cooperation via spectrum sharing for physical layer security in device-to-device communications underlaying cellular networks. *IEEE Transactions on Wireless Communications*, 15(8), p. 5651–5663.

Zhu, D., A.L. Swindlehurst, S.A.A. Fakoorian, W. Xu, and C. Zhao. (2014), Device-to-device communications: the physical layer security advantage. In: '2014 IEEE International Conference on Acoustics, Speech and Signal Processing (ICASSP), p. 1606–1610.

10

Automated Diagnosis of Skin Cancer for Healthcare: Highlights and Procedures

Maram A. Wahba and Amira S. Ashour

Department of Electronics and Electrical Communications Engineering, Faculty of Engineering, Tanta University, Egypt

Computer-aided diagnostic systems have aided the development of healthcare by introducing dedicated systems based on artificial intelligence for the early diagnosis of several abnormalities, such as skin cancer. The workflow of the automated skin cancer diagnostic system has a set of consecutive processes, namely image acquisition and pre-processing, image segmentation, feature extraction/ selection, and classification. This chapter outlines the framework of the automated dermoscopy-based skin cancer diagnostic systems by highlighting the main approaches that are commonly followed throughout the different processes of the system. The different imaging modalities for screening and diagnosing skin lesions are compared in terms of their advantages and limitations to present the advantages of the widely-used dermoscopy and the potential of applying multimodal imaging-based classification systems. Then, the sub-processes of the image preprocessing stage are presented, including color contrast enhancement and artifacts removal. Also, the different segmentation methods, and an overview of their proposed techniques are presented. The various techniques for extracting descriptive features from the segmented lesion are presented, such as color, dimensional, and texture features. In addition, this chapter introduces the dermoscopic-based features which are derived from clinical rules, such as the ABCD rule, Menzies method and the 7-point checklist to diagnose the malignancy of a lesion. Subsequently, the feature selection process is also discussed, along with some of its techniques. Finally, the different classifiers that were adopted in the context of dermoscopy-based skin cancer classification are reported. Finally, the evaluation criteria of the classifier's performance are discussed.

10.1 Introduction

Computer-aided diagnosis (CAD) systems have been emerging fast in the last two decades. The main purpose of CAD is to assist medical specialists in interpreting medical data, such as signals, images or videos by using artificial intelligence-based dedicated computer systems to provide them with second opinions. Many studies have proposed CAD systems for different applications, such as the diagnosis of breast cancer, lung cancer, colon cancer,

Engineering and Technology for Healthcare, First Edition.
Edited by Muhammad Ali Imran, Rami Ghannam and Qammer H. Abbasi.

coronary artery disease, and Alzheimer's disease, which have shown that CAD systems can significantly improve diagnostic accuracy. Thus, CAD has significant impact on healthcare, especially on the diagnosis of abnormalities in their early stages, which may be overlooked by traditional diagnostic methods. Also, CAD aids in reducing the burden of increased workload on physicians and reducing the inter-variability between their decisions (Fujita et al. 2008).

One of most serious and widely-spread cancers which is also an indicator of other diseases is skin cancer. It is caused by the deoxyribonucleic acid (DNA) mutations of skin cells leading to abnormal cells formation. These cells can grow at an irrepressible rate and develop cancerous masses. In addition, they may spread and invade other body parts (Narayanan et al. 2010). One of the primary causes of this epidemic is the frequent exposure to ultraviolet (UV) radiations either from sunlight or artificial sources as tanning beds. In addition, fair-skinned populations are the highest-risk individuals due to the lack of the melanin pigment in their skin (Narayanan et al. 2010). Other potential factors that increase the skin cancer risk include family history of skin cancer, rare genetic syndromes, medical conditions, and the exposure to toxic substances (Mayer and Goldman 2016).

Skin lesions are categorized mainly into benign (non-cancerous) and malignant (cancerous). Some of the most widespread benign lesions are nevus, seborrheic keratosis, dermatosis papulosa nigra, cherry angiomas, dermatofibromas, and solar lentigo. Malignant lesions include melanoma, basal cell carcinoma (BCC), and squamous cell carcinoma (SCC). The early diagnosis of skin cancer has a major impact in increasing recovery chances and survival rates. As an example, statistics gathered in 2017 have specified that the five year survival rate, which indicates the percentage of people who may live for at least five years after cancer diagnosis, is 92% if melanoma is early detected (Cokkinides et al. 2010).

This motivated the development of computer-aided skin lesion diagnosis systems, which are based on applying image processing for detecting and classifying skin lesions. These automated systems improve the healthcare system, as they overcome the low accuracy and infection drawbacks of the traditional diagnostic methods, namely the visual inspection and biopsies, respectively. Thus, these automated systems introduce a non-invasive artificial-intelligence based solution, which is highly accurate, safe, convenient, and accessible by both the patients and physicians, which supports the healthcare.

10.2 Framework of Computer-Aided Skin Cancer Classification Systems

In this section, we highlight the different approaches followed for the processes of the skin cancer CAD system, namely the image acquision process, image preprocessing, segmentation, feature extraction and selection, and classification. Also, the different metrics used for evaluating the classification performance are discussed.

10.2.1 Image Acquisition

There are several imaging techniques and modalities that are used for imaging and screening skin lesions (Ruocco et al. 2004), (Smith and MacNeil 2011), (Wassef and Rao 2013), which include digital photography as total cutaneous photography, multispectral imaging, laser-based enhanced diagnosis, such as optical coherence tomography (OCT), confocal scanning laser microscopy (CSLM), magnetic resonance imaging (MRI), ultrasound imaging, and dermoscopy. Among all the imaging modalities of skin lesions, dermoscopy is considered one of the most appropriate and practical imaging techniques in terms of image resolution, detailed visualization of the suspicious skin lesion, ease of use, and cost (Hibler et al. 2016). Moreover, combining different imaging technologies in a multimodal automated skin cancer classification system improves melanoma detection and other malignant lesions (Kawahara et al. 2018).

The dermoscopy, also called dermatoscope or epiluminescence microscopy (ELM), is a noninvasive skin imaging technique for in vivo morphological visualization of cutaneous skin structures on the superficial layers of the skin. It improves the physician's diagnosis based on the revealed morphological features. These features include shape, size, contours, structures, and color patterns that are hardly visible using the naked eye. The acquired dermoscopy images are usually impaired with artifacts, such as hairs and air bubbles that must be sufficiently reduced before further processing for improved recognition of the lesion under examination.

10.2.2 Image Pre-Processing

Several artifacts can severely impair the quality of the acquired images, which results in low diagnostic performance due to the difficulty in extracting the lesion (region of interest: ROI) from the background. Furthermore, the presence of false pixels, such as the pixels of hair and black frames, affects the diagnosis process. Generally, these artifacts can be categorized into: i) imaging-related artifacts, such as low contrast, fuzzy lesion borders, inconsistent illumination, black frames, dermoscopic gel, air bubbles, ruler, and ink markings; and ii) intrinsic dermal artifacts, including blood vessels, hairs, and skin texture. Thus, emerged the need of applying image enhancement processes for overcoming the effect of these artifacts without altering the main morphological features of the skin lesion by means of color contrast enhancement, and artifact removal.

10.2.2.1 Color Contrast Enhancement

Insufficient image contrast complicates lesion border detection in dermoscopy images. Two main techniques are commonly used to resolve such limitation, including the hardware-based approach, and the software-based approach (Wighton *et al.* 2011; Delalleau et al. 2011). Several algorithms and techniques have been proposed in this context. For example, Gómez et al. (2008) proposed the independent histogram pursuit (IHP) algorithm, which enhances the contrast between the lesion and the healthy skin in multi-spectral dermoscopy images. This algorithm estimates a set of image-dependent linear combinations using the image bands. For an N-band image, each combination corresponds to an N-dimensional line. The first combination is estimated as the component at which the image histogram is projected onto it would have maximal bimodal separation. This proposed linear transformation improves segmentation precision for multi-band dermoscopy images. For converting an RGB image to gray-scale, Celebi et al. (2009) determined the optimal weights that ensure the maximal histogram bimodal separability between the lesion and the background regions. Otsu's thresholding between-class variance was considered as a separability measure leading to more accurate region separation. An automated color calibration system is proposed by (Iyatomi et al. 2011), which compensates for any unintended color alterations that occurred through the imaging devices by applying modification filters to each dermoscopy image. The process obtained corresponding modified images from which different low-level color-related features are extracted. The inverse of the modification filter was applied for restoring the original images. Nevertheless, practically, these modifications are unknown to the restoration system. Thus, regression models that are based on the training data are built to estimate the relationship between the extracted features and the restoring procedure. These models can then be applied to restore any previously unknown image based on the extracted features' values. Other contrast enhancement techniques have been adopted, such as the adaptive histogram equalization (Pizer et al. 1987) and homomorphic filtering (Stockham 1972).

10.2.2.2 Artifact Removal

Several techniques have been adopted to remove artifacts from the dermoscopy images; one of the most straightforward methods is the application of smoothing filters, such as median, mean, or anisotropic diffusion filters. However, there are a few issues that should be considered upon using these filters, which are as follows (Celebi et al. 2015):

- Scalar and vector form filtering: these filters are designed to process single-channel scalar images such as the grayscale images. However, these filters may also be applied to multi-channel vector images by filtering each channel independently, then combining the results of all the filtered channels. This technique is called marginal filtering. Another solution is to apply vector filtering techniques (Celebi et al. 2007a), which process the image in its vector form directly.
- Mask size setting: the specified mask size is proportional to the amount of the smoothing resulting in the filtered image. Large mask size leads to blurring the lesion edges, while small mask size leads to noisy filtered images.. Thus, a reasonable strategy is to set the mask size in proportion to the image size (Emre Celebi et al. 2008).
- Computational time: regardless of the mask size, the median, mean, and Gaussian filtering algorithms require constant processing time. In contrast, for anisotropic diffusion filters (ADF), the mask size and the number of iterations have a direct impact on computational time.

Furthermore, a tailored technique can be applied to each type of artifacts as an alternative artifact removal method. The presence of hair, which is commonly occluding the lesion of interest, leads to segmentation errors, poor pattern-analysis, and low classification accuracy. For hair detection and removal, several studies were conducted, which involve three main parts, namely extracting long thin hair-like structures for constructing the hair mask, establishing an algorithm to discriminate and exclude the non-hair particles from the mask, and finally using the enhanced mask to guide the process of hair pixel value substitution without causing blurring, in addition to having a computationally efficient performance. Lee et al. (1997) proposed an automated hair detection and removal algorithm, which is known as the DullRazor algorithm; the methodology adopted in that study followed three main steps: i) determining the locations of dark hairs using morphological closing operations on the three color channels of the RGB dermoscopy image, ii) substituting the located hair pixels by nearby non-hair pixels using interpolation, and then, iii) smoothing the final result using adaptive median filtering. Schmid-Saugeona et al. (2003) adopted a hair removal method that was based on transforming the RGB image into the CIE L*u*v* color space, since the luminance component is the most appropriate for discriminating hairs from other pigmented structures. A morphological closing operation was applied on the luminance component, and then thresholding was applied to the difference between the original and the morphological closed image to obtain the hair mask. Consequently, the located hair pixels were replaced by their values in the morphological closed image.

Various hair removal studies have applied different techniques, such as applying closing-based top-hat filtering, followed by thresholding to detect the hair mask followed by partial differential equation (PDE) based inpainting for hair removal (Xie et al. 2009). Also, Fleming (Fleming et al. 1998) and Zhou (Zhou et al. 2008) used multi-constraint curvilinear structure detection for detecting hairs, and then applied exemplar-based inpainting for hair removal. Other proposed hair detection techniques, include using Radon transform (Jafari-Khouzani and Soltanian-Zadeh 2005) followed by using Prewitt operator for edge detection, employing a bank of directional filters (Barata et al. 2012), or applying morphological top-hat filtering on the channel which has the highest entropy which is followed by Otsu's thresholding (Afonso and Silveira 2012). However, Abbas (2013) introduced detecting the hair by combining the matched filter with the first-order derivative of the Gaussian (MF-FDOG), which was originally proposed for extracting the curve-like blood vessels in retinopathy images, along with morphological closing operations. Then, the hair-occluded regions were repaired by using fast-marching inpainting. These research efforts in the domain of dermoscopy image enhancement and preprocessing aim to facilitate the segmentation process, that is the lesion border detection, since refined border detection precision increases the fidelity of the extracted features. Subsequently, this will lead to improved classification and diagnosis accuracy.

10.2.3 Image Segmentation

The goal of using segmentation in automated skin lesion diagnostic systems is lesion border detection by extracting the lesion from the healthy background skin with high accuracy. Generally, the segmentation process depends on the similarity and discontinuity of the regions to be segmented in terms of specific characteristics, such as luminance, color, or texture. Discontinuities are detected at the abrupt changes in the intensity of the image pixels relative to their neighbors. Edge-based segmentation methods depend on the discontinuity criteria for detecting the image edges. However, other segmentation approaches are based on similarity criteria, which specify the sub-regions of a certain segment. Various segmentation approaches rely on similarity criteria to extract the skin lesions from the dermoscopy images, such as the thresholding-based, edge-based, and region-based segmentation approaches. In addition, other techniques based on active contours or artificial intelligence (AI) can be used for dermoscopic image segmentation (Masood and Ali Al-Jumaily 2013; Celebi et al. 2015; Oliveira et al. 2016; Ashour et al. 2018b; Guo et al. 2018; Dey et al. 2018; Ashour et al. 2018a; Ashour et al. 2019; Hawas et al. 2019; Ashour and Guo 2019).

10.2.3.1 Thresholding-Based Segmentation

Thresholding-based segmentation is one of the widely used techniques in skin lesion segmentation, which is based on exploiting the image histogram (Di Leo et al. 2010; Garnavi et al. 2011; Norton et al. 2012). The threshold levels separate the image into multiple regions based on the probability distribution of pixel intensities. Commonly, one threshold level is used for the extraction of the skin lesion from the surrounding skin. Different methods have been proposed for determining the optimal threshold value(s), such as Otsu's method (Otsu 1979), which is a global thresholding technique. Otsu's method selects the threshold value(s) that ensures the tightness of each separated class by minimizing its local variance, while maximizing the inter-class variance. Hence, it ensures the optimal separability between these classes. Nonetheless, Otsu's method performance degrades when the skin lesion smoothly fades into the surrounding skin. This may lead to some implications, such as that the segmented lesion may appear much smaller than its actual size, besides having distorted lesion edges.

Some researchers overcame these limitations by applying the adaptive thresholding technique (Silveira et al. 2009, Barata et al. 2014). Unlike Otsu's global thresholding technique, which applies the same threshold for all image pixels, adaptive thresholding calculates a threshold value for each image pixel to determine whether it belongs to the ROI or to the background. This compensates for the effects of non-uniform illumination over the image, by evaluating a small region of an image at a time, which is more likely to have consistent illumination. The threshold value can be calculated through two approaches, i) The Chow and Kaneko approach, or ii) local thresholding. In the Chow and Kaneko approach, the image is divided into an array of overlapping sub-images, and then the threshold of each sub-image is calculated based on statistical information. Eventually, the threshold for each pixel is evaluated by interpolating the results of the sub-images. Nevertheless, this approach is computationally expensive, and thus inconvenient in real-time applications. In the local thresholding approach, an individual threshold for each pixel is statistically calculated based on the intensity values in its local neighborhood. The size of neighborhood should be chosen large enough to include sufficient foreground and background pixels, or else a poor threshold will be chosen. In contrast, selecting a neighborhood that is too large may disrupt the uniform illumination assumption leading to the same drawbacks of global thresholding.

10.2.3.2 Edge-Based Segmentation

The edge-based segmentation techniques in Pires and Barcelos (2007), Barcelos and Pires (2009), and Abbas et al. (2012) are based on identifying the image discontinuities for detecting these edges, which is achieved through two main steps: i) the edge pixels are identified using one of several edge detection operators; then, ii) the post-processing or edge-linking process. In this process, the identified adjacent edge pixels are combined into edge chains to form the region border. Thus, other superior edge-linking methods have been adopted, such as Hough transform, although it does have higher complexity and sensitivity to noise, but it can considerably handle the edge gaps problem. This canny technique has enhanced performance in edge-detection, while suppressing the speckle noise at the same time by using a low-pass gaussian filter. Then, it applies edge detection, using techniques such as Sobel, Prewitt, and so on to obtain the gradient magnitude image, which is then applied to thresholding.

10.2.3.3 Region-Based Segmentation

The objective of the region-based techniques is to fulfill two requirements, namely, to produce regions that are as large as possible, and to produce coherent (uniform) regions while allowing some flexibility for variation within the region. However, a tradeoff is present between these two requirements, such that applying a strict similarity measure between the pixels will lead to great coherency. In contrast, this may cause over-segmentation, which means that a region will be much smaller than the actual object. However, increasing flexibility may cause a region to cross an object boundary and leak to other objects. Thus, different algorithms have been developed aiming for an optimal solution to this tradeoff.

The developed algorithms can be categorized into one of three criteria, which are merging algorithms, splitting algorithms or splitting and merging algorithms. Generally, the region merging (Emre Celebi et al. 2008; Nock and Nielsen 2004; Celebi et al. 2007b; Iyatomi et al. 2008a; Lissner and Urban 2012) starts by selecting a suitable seed, then merges adjacent uniform areas until it forms coherent regions. This recursive process stops when no further merging is possible. If the starting seed is a single pixel, the algorithm will merge the candidate neighboring pixels to it according to one of these approaches. The first approach entails comparing each candidate pixel to the seed through evaluating their similarity in terms of a specified function. However, this approach is sensitive to the choice of the seed pixel and may lead to over-segmentation. The second approach reduced the dependency on the seed pixel choice by comparing each candidate pixel to the last merged pixel instead of to the seed itself. In contrast, if the transition is gradual enough, it may cause significant drift while the algorithm starts growing further from the original seed. For example, adjacent objects that collectively have gradual color gradient can be mistakenly considered as a single region. The third approach solved this problem by comparing each candidate pixel to the aggregate statistics of the merged region. However, some algorithms are based on choosing multiple seeds to better describe the region statistics. Wong et al. (2011) proposed a method for segmenting dermoscopy images using iterative stochastic region merging (Nock and Nielsen 2004). It was based on a probability function of the merged regions. This algorithm has proven to be robust to image artifacts, and to perform successfully in the presence of varying illuminations and in cases of low contrast between the lesion and the surrounding skin around the lesion boundary.

Another region-based segmentation criterion is region splitting, which starts splitting the whole non-uniform image recursively until smaller uniform regions are formed. Its major issue is determining where to perform the partitioning. However, in most cases the splitting is used as the initial stage in the splitting and merging algorithms. Splitting and merging algorithms are computationally less expensive than the merging algorithms (Gao et al. 1998; Silveira et al. 2009). These algorithms split the image into smaller and smaller regions until each individual region is coherent; then it recursively merges those regions to produce larger coherent regions.

10.2.3.4 Active Contours-Based Segmentation

Several active contour-based segmentation algorithms have been developed and applied for dermoscopy images (Silveira et al. 2009; Zhou et al. 2010; Zhou et al. 2013; Ma and Tavares 2016; Nagieb et al. 2018). Active contours are computer-generated curves that tend to delineate an object outline in a given image. These curves are initially placed in the image and undergo a deformation process until they find their optimal position on the object's boundary. This deformation process is driven by energy forces. The internal energy force defines the intrinsic shape properties, such as the curve rigidity and elasticity, thus, preserving its smoothness during deformation. However, the external energy force drives the curve to the desired boundary. In addition, image-based energy drives the curve toward interesting image features. Different techniques have been developed for the deformable model, which can be classified as either parametric (Zhou et al. 2010; Zhou et al. 2013) or geometric (Chan et al. 2000; Ma and Tavares 2016), according to the technique used for tracking the curve movement. Snake models and gradient vector flow (GVF) models are examples for the parametric models, but these models have limitations including the difficulty in converging at boundaries that have large curvatures, also the initial curve must be near the object's boundary; besides the stopping criterion depends on the image gradient, which in some cases is not high enough, thus causing bad detection. Unlike the parametric models, the geometric models are less dependent on the initial curve conditions and are more adaptive to the topological changes during the curve evolution.

Chan, Sandberg, and Vese (2000) proposed an active contour model without edges which is based on the average of the image pixels intensities instead of the image gradient. Thus, this model applies the concepts of the Mumford-Shah (Mumford and Shah 1989) and level set (Ma and Tavares 2016) segmentation techniques. This model introduced a "fitting" term for energy minimization, which is calculated by means of energy functions, for identifying whether the object of interest is inside or outside the curve. This model was used in the segmentation of skin lesions due to its superiority over the other active contour-based segmentation models in terms of providing effective boundaries detection even at noisy images without the need of applying previous denoising, besides the flexibility in defining the initial curve. Also, unlike the traditional active contour-based models, the object detection is possible even in the presence of very smooth boundaries and varying intensities.

10.2.3.5 Artificial Intelligence-Based Segmentation

Several algorithms have been proposed for segmenting skin lesions based on artificial intelligence, in which an image pixel is classified as either being at the ROI or at the image background. These techniques include neural networks, evolutionary computation, and fuzzy logic, which can be used as standalone segmentation techniques, or combined with other segmentation methods for enhancing their performance. The artificial neural networks (ANNs) are layered networks that are composed of interconnected processors which apply algorithms, attempting to emulate the skills of a human brain, such as identifying and segmenting images for example. ANNs have been applied for segmenting skin lesion images, for example Vennila et al. (2012) have proposed using ANNs for extracting the ROI of the melanoma dermoscopy image using machine learning algorithms. Three different algorithms have been examined, which are the radial basis function network (RBF), the back propagation network (BPN), and the extreme machine learning (ELM). First, the ANN was trained using extracted features from dermoscopy images that were manually outlined by physicians. After that learning phase, the ANN was able to segment the input images. Reportedly, the ELM algorithm had the best performance in terms of accuracy and segmentation results, besides its faster training period compared to the RBF and BPN neural networks. The performance of traditional ANNs has been reportedly improved by using the genetic algorithm (GA), which is a searching and optimization technique. GA can search for the optimal solution through complex datasets, thus it can optimize the ANNs' parameters, such as the weights, the thresholds and the learning rate, thus accelerating the training speed and improving its performance. For example, Jianli and Baoqi (2009) have proposed improving the performance and convergence speed of the standard back propagation (BP) neural network through optimization using the GA. As the structure and learning regulation of that network is simple enough, the researchers focused only on optimizing the weights and thresholds to improve its training and simulation speed, besides its ability to handle the large grayscale sample set. The optimized BP network achieved faster segmentation speed; also, the segmented edges had clear contours and continuous edges.

Fuzzy-based techniques are considered as a sub-branch of the artificial-intelligence-based methods. Fuzzy logic is a computing approach that deals with the uncertainty of the given data, unlike common "true or false" logic. Many algorithms have proposed applying this concept in skin lesion image segmentation (Maeda et al. 2007; Maeda et al. 2008; Yüksel and Borlu 2009; Khan et al. 2009; Silveira et al. 2009; Masood and Al-Jumaily 2013). The fuzzy methods are combined with the traditional segmentation methods, such as thresholding-based, clustering-based, and region-based methods, for improving their performance. For example, Maeda et al. (2008) and Silveira et al. (2009) combined the fuzzy method with splitting and merging algorithms for segmenting dermoscopy images. Also, fuzzy methods have been used to improve the threshold selection in thresholding-based segmentation (Yüksel and Borlu 2009). In addition, the neuro-fuzzy approach (Castiello et al. 2004) introduced fuzzy logic to the neural networks for segmenting dermatoscopic images. Even clustering-based segmentation methods have been improved by the fuzzy methods, such as the fuzzy c-means (FCM) clustering algorithm (Masood and Al-Jumaily 2013) in which the fuzzy methods have added more flexibility in finding the clusters' centers using partial membership.

The dermoscopy image segmentation domain is still undergoing continous research for reaching optimal segmentation performance. For the performance evaluation of the automated border detection algorithms, the resulting images are compared against the borders that are manually detected by dermatologists. These validating images are commonly referred to as the gold-standard or the the ground truth images, which are used for both subjective and objective evaluation (Celebi et al. 2015). Subjective evaluation indicates the visual evaluation of the segmentation results by dermatologists. However, objective evaluation involves using the validating ground truth images for quantifing the border detection errors. These quantitative measures are commonly based on the confusion matrix concepts of true/false positives/negatives, such as the sensitivty, specificity, accuracy, precision, recall, and Jaccard index. The significance of the segmentation process is based on its major impact on the subsequent feature extraction process, at which different features including the border shape are used to discriminate the different classes of skin lesions. So, the faulty presence of background skin within the segmented region would lead to inaccurate classification results.

10.2.4 Feature Extraction

Several feature extraction approaches have been employed in the computational analysis of dermoscopy images; mostly these approaches are based on the visual diagnosis procedures that are clinically used by dermatologists. Thus, the computer-aided feature extraction approaches aim to identify the different lesion types by extracting descriptive features from the segmented lesions. Various types of features have been proposed for identifying these lesions, such as color-based features, dimensional features, texture-based features, and dermoscopic features.

10.2.4.1 Color-based Features

Color features are commonly extracted through statistical calculations over the different color channels. The statistical features (Chang et al. 2005; Celebi et al. 2007b; Iyatomi et al. 2008b; Iyatomi et al. 2008a; Alcón et al. 2009; Maglogiannis and Doukas 2009), such as the mean, the variance, the standard deviation, and the skewness can be computed from the color channels of the different color spaces. Additionally, these features can be extracted from precise regions, such as those associated with the lesion's border, namely the lesion's inner and outer peripherals, so that the malignancy of the lesion could be determined in case of having a steep pigment transition along these peripherals. Different techniques have been applied to locate the lesion's peripheral regions such as estimating a circular region centered at the lesion's centroid (Chang et al. 2005), or by using recursive erosion process (Iyatomi et al. 2008b; Iyatomi et al. 2008a). Also, other techniques applied fast Euclidean distance transform algorithm (Celebi et al. 2007b).

10.2.4.2 Dimensional Features

Several descriptors (Claridge et al. 1992; Andreassi et al. 1999; Lee et al. 2003; Maglogiannis et al. 2006; Celebi et al. 2007b) have been proposed for measuring these features. Some studies have used simple dimensional descriptors, such as the lesion's perimeter and area. Nevertheless, several descriptors have been used for extracting shape features, such as aspect ratio, asymmetry index, bulkiness score, circularity factor, equivalent diameter, solidity, irregularity index, fractality of borders, and compactness. Several studies have concluded that the border shape descriptors, such as fractal dimension and edge abruptness (Chang et al. 2005; Celebi et al. 2007b; Clawson et al. 2009; Ramlakhan and Shang 2011; Garnavi et al. 2012) are highly efficient in detecting malignant melanoma, whether using the computational analysis methods or even clinically.

10.2.4.3 Texture-Based Features

The image texture is the regularity of certain patterns repeated in a local region in the image, which provides the potential of distinguishing between different image classes, such as classifying skin lesion dermoscopy images. The texture features include statistical methods, filtering-based methods, spatial-based methods, and model-based methods. Statistical methods include the Grey-level co-occurrence matrix (GLCM), neighboring gray level dependence matrix (NGLDM), and run length matrix (Maglogiannis et al. 2006; Celebi et al. 2007b; Nie 2011; Sheha et al. 2012). While, filtering-based methods include Gabor filter banks (Yuan et al. 2006), laws masks, and wavelet transforms (Surowka 2008; Garnavi et al. 2012). Spatial-based methods rely on calculating the histogram of occurrences for a certain parameter, such as histogram of gradients and histogram of lines (Barata et al. 2014). The model-based methods attempt to match the texture of the image by means of mathematical models, such as autoregressive models (Yuan et al. 2006).

10.2.4.4 Dermoscopic Rules and Methods

Several automated skin lesion analysis systems and studies are based on the clinical diagnostic methods and rules that are commonly applied by the dermatologists in the non-invasive evaluation of the pigmented skin lesions. These clinical methods include the ABCD rule, Menzies method, 7-point checklist, and pattern analysis. Each of these rules and methods consist of a set of features that are clinically evaluated in a visual manner by the dermatologist to determine the lesion's malignancy based on the presence or absence of these features. These features can be categorized into shape features, such as asymmetry and diameter, border irregularity features, color-based features, and high-level dermoscopic features such as streaks, globules, dots, blue-white veil, and so on. Thus, several descriptors have been proposed to characterize these features quantitively through computational analysis, which improves the clinical diagnosis accuracy and reduces the probability of visual misclassification.

ABCD Rule

The ABCD rule differentiates between malignant melanomas and the benign nevus through main five criteria which are the asymmetry (A), the border irregularity (B), the color variations (C), and the diameter or differential structures (D). This rule is used in the clinical analysis as shown in Table 10.1 and in the computer-based analysis of dermoscopy images as shown in Table 10.2.

Table 10.1 shows that all the five aforementioned criteria are used for clinical analysis, while Table 10.2 shows that for computational analysis without considering the elevation and evolution features. This is due to the difficulty of having a dataset with at least two images for the same lesion that are captured over a sufficient time interval to assess its

Table 10.1 Clinical ABCDE Rule Criteria for Diagnosing Malignant Melanoma apart from Benign Nevus

Clinical Analysis ABCDE rule		
Feature	**For malignant melanoma**	**For benign nevus**
Asymmetry (A)	Unsymmetrical (i.e., if a line is drawn to divide the lesion, the two halves will not match)	Symmetric (i.e., if a line is drawn to divide the lesion, the two halves will match)
Border (B)	The border tends to be irregular, edges may be notched or scalloped	Smooth even borders
Color (C)	Different color shades of brown, black, and sometimes red or blue	One color, often a single shade of brown
Diameter (D)	Larger diameter than benign moles (about 6 mm or more)	Usually have smaller diameter than malignant ones
Evolution (E)	Change in size, shape, or color	No change
Elevation (E)	Higher than skin surface	Smooth on skin surface

Table 10.2 ABCD Score Calculation for Diagnosing Malignant Melanoma

Image-based dermoscopy analysis ABCD rule			
Feature	**Definition**	**Score**	**Weight factor**
Asymmetry (A)	Lesion asymmetry by 0,1, or 2 orthogonal axes	0–2	1.3
Border (B)	Border abruptness based on the number of segments having irregular borders after dividing the lesion into 8 wedge-shaped segments	0–8	0.1
Color (C)	The presence of six possible colors (black, red, dark-brown, light-brown, blue-grey, and white)	1–6	0.5
Diameter (D)	Diameter more than 6 mm is assigned a score of 5	1–5	0.5

evolution over time, and the need of designing a three-dimensional digital imaging system to capture the elevation information.

Each of these features in Table 10.2 are assigned a certain score accordingly, then the total dermoscopy score (TDS) is calculated from these scores by the following equation:

$$TDS = A \times 1.3 + B \times 0.1 + C \times 0.5 + D \times 0.5 \tag{10.1}$$

Such that if the TDS is less than 4.75, it indicates benign lesion, while a TDS in the range of 4.8 to 5.45 presents some potential of suspicious melanoma; however, a TDS that is greater than 5.45 specifies a highly suspicious melanoma.

The asymmetry of a lesion is commonly calculated by determining the lesion's centroid, then dividing the lesion along each of the horizontal and vertical axes of symmetry into two regions. Accordingly, the corresponding regions are overlapped to find their symmetry. The axis of symmetry can be defined based on the principal axis of inertia (Chang et al. 2005), the shortest or longest diameter (Ng et al. 2005), minor and major axis orientation (Celebi et al. 2007b; Garnavi et al. 2012). Several techniques can be used to calculate the asymmetry index, such as the simple application of the XOR operation between the two sub-regions to estimate the difference between them (Chang et al. 2005). Moreover, some studies have proposed obtaining other asymmetry features, which allow the assessment of the color, shape and border asymmetries. Celebi et al. (2007b) have considered evaluating the color asymmetry to detect the skin lesion malignancy. The shape dimensional measures include the irregularity index (Garnavi et al. 2012), convex hull (Alcón et al. 2009), solidity (Celebi et al. 2007b; Alcón et al. 2009), which is defined as the ratio between the area of the lesion and its convex hull, aspect ratio (Celebi et al. 2007b; Alcón et al. 2009), circularity index (Iyatomi et al. 2008a; Alcón et al. 2009; Iyatomi et al. 2010; Garnavi et al. 2012), compactness (Celebi et al. 2007b), rectangularity (Celebi et al. 2007b; Alcón et al. 2009), entropy measures, greatest and shortest diameter (Garnavi et al. 2011), perimeter (Alcón et al. 2009; Garnavi et al. 2012), and the lesion area.

The lesion border irregularity is a sign that indicates its potential malignancy. Several methods have been proposed for evaluating the border's regularity, such as the estimation of the mean and variance of the associated edges gradient (Alcón et al. 2009), the compactness index, fractal dimension. For example, Iyatomi et al. (2008a) separated the lesion into eight equiangular regions. Afterward, the gradient of the color intensity, and the color intensity ratio between the inner and outer areas of the lesion were computed. Also, Clawson et al. (2009) have proposed the harmonic wavelet transform descriptor, which represents the border energy distribution over a multi-scale of roughness to analyze the border irregularity. Garnavi et al. (2012) developed a boundary-series model by determining the distance between each border pixel and the lesion's centroid. Then, the developed boundary series was analyzed statistically in the frequency domain by applying a three-level wavelet transform, and also spatially using the histogram of the boundary-series.

The non-uniform color of the skin lesion is one of the significant indicators in identifying its malignancy. In the automated analysis of dermoscopy images, the number of colors that are present is counted out of six possible colors, which are black, dark-brown, red, light-brown, white, and blue-grey. Skin lesions are commonly in elliptical form; thus, a lesion's diameter is usually estimated as the greatest distance between any two points on a lesion's border (Abbasi et al. 2004), which is calculated as the semi-major axis of the best-fit ellipse (She et al. 2007).

Menzies Method

The Menzies method uses "negative" and "positive" features for melanoma diagnosis. The negative features include the presence of a single color within the lesion's area, and the symmetrical pigmentation pattern. However, the positive features include the presence of blue-white veil, pseudopods, multiple brown dots, scar-like depigmentation, radial streaming, peripheral black dots/globules, multiple (five to six) colors, multiple blue-gray dots, and broadened pigment network (Johr, 2002). None of the two "negative features" should be found, and at least 1 of the 9 "positive features" must be present for a lesion to be classified as melanoma. On the other hand, benign lesions have a single colored, symmetric pigmentation pattern. Some studies applied this method for automated skin lesion classification. For example, Silva and Marcal (2013) proposed estimating the presence of the six colors (white, red, light-brown, dark-brown, blue-grey, and black) that are associated with the lesion's malignancy according to the Menzies method. The transformed divergence and Jeffries-Matusita separability measures were applied to estimate the degree of separability between the color classes.

7-Point Checklist

The 7-point checklist provides a reliable method for diagnosing the melanoma based on assigning scores to the suspicious lesion based on two criteria, namely the major and minor criteria. The major criteria aim to detect the atypical pigment networks, gray-blue areas, and atypical vascular patterns. Each of those features if detected represents a score of 2. Then, the minor criteria give a score of 1 to any of the subsequent features, which are radial streaming (streaks), blotches, irregular dots and globules, and regression patterns. A minimum score of 3 is enough to classify the lesion as melanoma. Numerous studies (Betta et al. 2005; Celebi et al. 2008; Di Leo et al. 2010) have focused on the detection of one or multiple features of the 7-point checklist. For example, Celebi et al. (2008) have used texture and color features in order to detect the blue-white veil area. The texture features were calculated based on the GLCM. In addition, the color features were estimated based on the concept of the relative and absolute color, which were calculated according to the relative RGB ratio, normalized relative RGB ratio, relative RGB difference, normalized relative RGB difference, and chromaticity coordinates. For detecting the occurrence of two criteria of the 7-point checklist, Betta et al. (2005) proposed an algorithm based on texture features estimated chromatic and shape parameters. First, the irregular streaks were identified at the sub-images at which brown pigmentation have coincided with irregular contours. Thus, the original image was divided into 16 sub-images at which the border's irregularity ratio, as well as the presence of brown color were measured at each of them and compared to a pre-defined threshold. However, the atypical pigment network was identified according to the analysis of the spectral content which is obtained by applying the fast Fourier transform (FFT). According to the Johr's conclusion (Johr 2002), the automatic extraction of features that are applied in the ABCD rule is computationally less expensive than the ones that are applied in the Menzies method, or the 7-point checklist. Moreover, the ABCD rule has high reliability in clinical diagnosis. So, most of the automated computer-aided diagnostic systems rely on the ABCD rule as the basis of their feature extraction step.

However, it can be deduced from the research on the development of computer-aided skin lesion diagnostic systems that a combination of different feature extraction criteria is crucial for the early detection of malignant melanoma and other types of skin lesions. On the other hand, the high-dimensional feature space may complicate the classification process. Consequently, it is essential to define the features that are the most significant and exclude the redundancies.

10.2.5 Feature Selection

Various feature selection techniques were proposed for the feature sub-set selection in the automated skin lesion diagnostic systems including best-first search strategies (Maglogiannis and Doukas 2009), greedy forward selection, linear forward selection (Chang et al. 2005; Iyatomi et al. 2008a; Iyatomi et al. 2010), filter-based methods, and principle component analysis (PCA). For feature sub-set evaluation, several filter-based methods were proposed including, relief, mutual information-based feature selection (MIFS) (Celebi et al. 2007b), and correlation-based feature selection (CFS) (Garnavi et al. 2012). Another approach for feature selection is to consider it as an optimization problem using heuristic strategies, greedy search, or genetic algorithms (Handels et al. 1999). Roß et al. (1995) reduced the 87 extracted features representing the surface profiles of the lesion into the 5 most significant features using sequential forward selection algorithm. Ganster et al. (2001) applied sequential forward floating selection (SFFS) and sequential backward floating selection (SBFS). The results shown that the best selection performance was achieved using 10 to 15 features, and it reduced by selecting 20 features. Celebi et al. (2007b) used the area under curve (AUC) to evaluate the performance of the CFS feature selection algorithm, which achieved the best performance using the highest 18 features. Rohrer et al. (1998) performed a study which established that a small-sized feature sub-set had better performance than using large feature sub-sets, which was validated by the Ruiz study (2011) that evaluated the SFFS and SBFS showing that the minimum error rate was achieved with a sub-set of 6 features. However, a substantial rise in the classification error rate was observed using a sub-set of more than 20 features.

10.2.6 Classification

Various machine learning algorithms have been applied in the classification systems of dermoscopy images, such as the nearest neighbors (Rahman et al. 2008; Barata et al. 2014) decision trees (Chang et al. 2005; Celebi et al. 2008; Clawson et al. 2009; Garnavi et al. 2012), Bayesian learning (Maglogiannis and Doukas 2009; Garnavi et al. 2012), artificial neural networks (Iyatomi et al. 2008a; Clawson et al. 2009; Maglogiannis and Doukas 2009; Silva and Marcal 2013) and SVMs (Celebi et al. 2007b; Rahman et al. 2008; Maglogiannis and Doukas 2009; Garnavi et al. 2012; Barata et al. 2014). Nearest neighbors classifiers are characterized by their simple implementation and their ability to deal with correlated features. They determine the class of a testing sample by evaluating the similarity between it and each training sample. Thus, they are also called lazy algorithms because they tend to use all (or most) of the training samples for the testing purpose; besides they have no explicit training phase. However, this implies that they are computationally expensive as they need to store all of the training samples. Also, the prediction time decreases with large datasets. Moreover, it is sensitive to the presence of irrelevant features. One of its common algorithms is the KNN algorithm, which classifies the data based on a majority vote of its k-nearest neighbors, such that K is usually an odd number that represents the number of neighbors. For example, Barata et al. (2014) conducted a comparative study of several techniques such as Euclidean, Kullback–Leibler, and Kolmogorov in order to measure the distance to the nearest neighbors for different K values. However, the authors concluded that it is unfeasible to clarify an optimal technique among them because the evaluation results are clearly dependent on the testing situation.

Decision tree (Han et al. 2011) algorithms are one of the simplest supervised learning methods in both understanding and implementation. This model resembles an upside-down tree such that the starting point is at its roots and the final decision is at its leaves. First, training data is used to create the decision tree by the following means: the best feature of the dataset (i.e., according to information gain or Gini index) is considered as the root node which splits the remaining training set into semi-equal sub-sets. Each sub-set has the same value of that feature. This step is repeated while moving deeply in the model by considering other features at each branch that further splits the training set until the model reaches the final decisions on its leaf nodes. This model is quite straightforward and is easily generalized over unseen data. However, one of the major drawbacks of decision trees is its high over-fitting potential in case of having many branches due to data irregularities and outliers. There are also difficulties in dealing with correlated features. Various algorithms based on decision trees have been applied in skin lesions classification such as classification and regression trees (CART) (Maglogiannis and Doukas 2009), Naïve Bayes/decision trees (NBTree) (Maglogiannis and Doukas 2009), decision tree learner (J48) (Alcón et al. 2009), C4.5 (Chang et al. 2005), C5.0 (Clawson et al. 2009), logistic model tree (LMT) (Garnavi et al. 2012), classification and regression (C&R)(Clawson et al. 2009), and random forest (RF) (Garnavi et al. 2012).

Bayesian methods evaluate the probability that a given set of features belongs to a certain class, assuming that these features are independent (Congdon 2007). Examples of Bayesian learning-based methods applied in skin lesions classification include the Bayes networks (Alcón et al. 2009; Maglogiannis and Doukas 2009), Hidden Naïve Bayes (HNB) (Maglogiannis et al. 2006), and Naïve Bayes multinomial (NBL)(Garnavi et al. 2012). Although Bayesian methods require only a small portion of training data to compute the parameters necessary for classification, they assume that the features need to be independent.

The ANNs are neural-inspired systems which are intended to imitate how humans learn by using an interconnected network of nodes similar to the vast network of neurons in a brain (Haykin 1994). These nodes form a parallel distributed system composed of input and output layers which are interconnected by weighted links. As a supervised machine learning model, the weights are adjusted during the learning phase to predict the corresponding class based on the input samples. The multilayer perceptron (MLP) is one of the most common architectures of ANNs (Iyatomi et al. 2008a) since such a network has high capability and flexibility in solving several non-separable problems. ANNs have been applied in solving many skin lesion classification problems; however, their disadvantages include their long training time, the dependency of the network's complexity on the data dimensionality, and its high potential of over-fitting.

SVMs are widely applied to classify skin lesions due to their good generalization properties (Steinwart and Christmann 2008). They apply statistical learning-based methods to select a hyper-plane that best separates the classes of the input data. In most cases, the input feature space is not linearly separable. The SVM has a special ability to separate nonlinear separable samples by using kernel functions. The kernel function transforms the low dimensional feature space into a higher dimensional space. Hence, converting the non-separable problem into a separable problem. Kernel functions include: linear kernels, polynomial kernels (Yuan et al. 2006; Wahba et al. 2017; Wahba et al. 2018) as the quadratic and cubic function kernels and, gaussian kernels as radial basis function (RBF) (Celebi et al. 2007b; Garnavi et al. 2012; Barata et al. 2014) while it is also possible to develop other forms of kernels. Typically, the SVMs achieve the best classification results and the highest accuracy especially in high dimension data sets; however, it is sensitive to noise.

10.2.7 Classification Performance Evaluation

The classification performance metrics are computed based on the number of accurately identified images, and the misidentified images. Accordingly, four parameters can be obtained, namely, the true positive (TP), the false positive (FP), the true negative (TN), and the false negative (FN), as shown by the confusion matrix in Figure 10.1.

Subsequently, these four parameters are used to evaluate the classifier's performance metrics. Some of the widely applied performance metrics are reported in Table 10.3.

Moreover, the receiver operating characteristics (ROC) curve (Fawcett, 2004) is an important two-dimensional graphical representation of the classifier's performance. This curve represents the relation between the sensitivity (TPR), which is plotted on the Y-axis versus the false positive rate (FPR) that represents the complement of specificity on the X-axis for several threshold settings. The area under curve (AUC) is commonly measured from the ROC curve which represents the classifier's expected predictive performance as a single

Figure 10.1 Confusion matrix of binary classification model

		Predicted class P	Predicted class N
Actual class	P	True positives (TP)	False negatives (FN)
Actual class	N	False positives (FP)	True negatives (TN)

Table 10.3 Some of the commonly used metrics in evaluating the classifier's performance

$$Accuracy = \frac{TP + TN}{TP + TN + FP + FN} \times 100\%$$

$$Diagnostic\ Accuracy = \frac{TP}{TP + FP + FN}$$

$$Sensitiviy = recall = \frac{TP}{TP + FN} \times 100\%$$

$$Specificty = \frac{TN}{TN + FP} \times 100\%$$

$$F - measure = \frac{2TP}{2TP + FP + FN} \times 100\%$$

$$Positive\ predictive\ value\ (PPV) = Precision = \frac{TP}{TP + FP} \times 100\%$$

$$Negative\ predictive\ value\ (NPV) = \frac{TN}{TN + FN} \times 100\%$$

$$Error\ probability = \frac{FP + FN}{TP + TN + FP + FN} \times 100\%$$

scalar value in the range between 0 and 1. Thus, the increase in the AUC value indicates better classification performance, such that an AUC value equals to 1 indicating a perfect classification performance.

10.2.8 Computer-Aided Diagnosis Systems in Dermoscopic Images

The performance of the classification process is dependent on the outcome of the previous processes, as shown in Figure 10.2, which demonstrates the block diagram of the dermoscopy image CAD system. The output of the skin lesion classification may be binary, such as evaluating the lesion's malignancy or classifying it as either melanocytic or non-melanocytic. Also, it can be extended to discriminate between the different skin lesion classes. For example, Shimizu et al. (2015) have introduced a decomposition classification model to distinguish between four lesion classes, namely melanoma, nevus, BCC, and seborrheic keratosis. The results reported detection rates of 82.51%, 82.61%, 80.61% and 90.5% for nevi, BCC, seborrheic keratosis and melanoma, respectively which outperformed the conventional flat model classifier. However, Celebi et al. (2007b) applied the SVM to distinguish between melanomas and benign lesions; however, Iyatomi et al. (2008a) employed the ANN classifier, which was then compared to the performance of other classifiers. In the study of Surówka (2008), the SVM (RBF kernel) achieved better performance than the ANN. Moreover, Clawson et al. (2009) have also compared the performance of

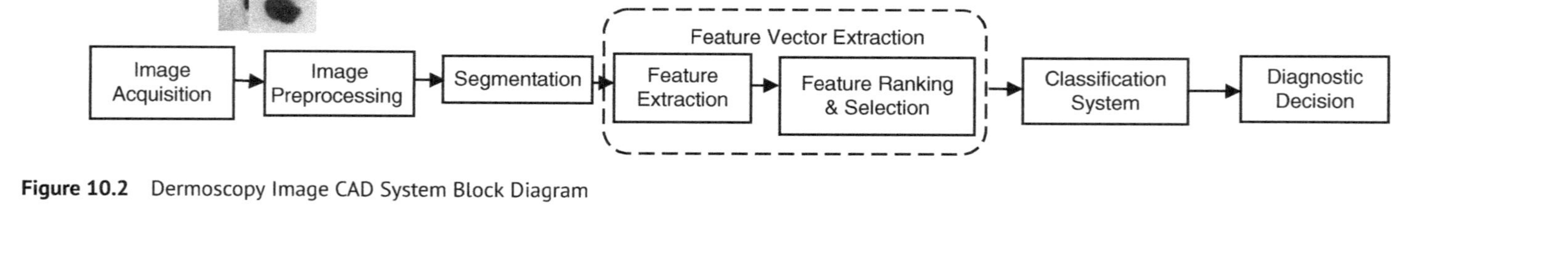

Figure 10.2 Dermoscopy Image CAD System Block Diagram

ANNs to other classifiers such as C&R and C5.0 in which the C5.0-classifier achieved the best performance.

On the other hand. Maglogiannis and Doukas (2009) compared the performance in terms of accuracy, root means square (RMS), TP and FP rates, and AUC of the following classifiers: ANN with multilayer perceptron, SVM, Bayesian networks, NBTree, CART, RBF networks, NBL, multinomial logistic regression (MLR), KStar, locally weighted learning (LWL), and classification via regression. The authors classified different skin lesion classes by performing three experiments. First, the classification problem between melanoma and nevus classes was considered in which MLR, LWL, SVM, and CART provided the best performance. Then, the classification between dysplastic and non-dysplastic lesions was considered. In that case, SVMs, MLPs, and Bayes networks performed better than other classifiers according to the corresponding ROC curves. The last experiment was based on the classification between melanotic, dysplastic, and non-dysplastic lesions in which the SVM had the best performance.

10.3 Conclusion

Computer-aided diagnosis of skin lesions is one of the vast applications of aritifical intelligence in the healthcare domain. Several research efforts have been invested in the various processes of dermoscopy-based CAD systems. The dermoscopy image is pre-processed by color contrast enhancement, and artifact removal to overcome the imaging-related artifacts, and the intrinsic dermal features. Different approaches have been proposed for lesion segmentation including thresholding-based, edge-based, and region-based segmentation approaches. In addition, other techniques based on active contours or artificial intelligence (AI). Several features have been proposed for quantifying the lesion, which include color, dimensional, and texture features; in addition to dermoscopic-based features which are derived from clinical rules, such as the ABCD rule and Menzies method. The significance of the extracted features is assessed by the feature selection methods aiming to decrease intra-class variance and increase the inter-class separation, which improves the classification diagnostic decision. Various machine learning algorithms have been applied in the classification systems of dermoscopy images, such as the nearest neighbors, decision trees, Bayesian learning, ANNs, and SVMs.

Acknowledgment

This work is supported by Tanta University [Grant ID: TU: 19-03-01]. The authors are thankful to Tanta University for this funding.

References

Abbas, Q., M. Celebi, M., and I.F García. (2012). Skin tumor area extraction using an improved dynamic programming approach, *Skin Research and Technology*, 18, 133–142.

Abbas, Q., I.F. Garcia, M. Emre Celebi, and W. Ahmad. (2013). A feature-preserving hair removal algorithm for dermoscopy images, *Skin Research and Technology*, 19(1), p.27–36.

Abbasi, N. R., H.M. Shaw, D.S. Rigel, R.J. Friedman, W.H. Mccarthy, I. Osman, A.W. Kopf, and D. Polsky. (2004). Early diagnosis of cutaneous melanoma: revisiting the ABCD criteria, *Jama*, 292, p.2771–2776.

Afonso, A., and M. Silveira. (2012). M. Hair detection in dermoscopic images using percolation, Engineering in Medicine and Biology Society (EMBC), 2012 Annual International Conference of the IEEE, IEEE, p.4378–4381.

Alcón, J. F., C. Ciuhu, W. Ten Kate, A. Heinrich, N. Uzunbajakava, G. Krekels, D. Siem, et al. (2009). Automatic imaging system with decision support for inspection of pigmented skin lesions and melanoma diagnosis, *IEEE Journal of Selected Topics in Signal Processing*, 3, p.14–25.

Andreassi, L., R. Perotti, p.Rubegni, M. Burroni, G. Cevenini, M. Biagioli, p.Taddeucci, et al. (1999). Digital dermoscopy analysis for the differentiation of atypical nevi and early melanoma: a new quantitative semiology, *Archives of Dermatology*, 135, p.1459–1465.

Ashour, A.S., C. Du, Y. Guo, A.R. Hawas, Y. Lai, and F. Smarandache. (2019). A novel neutrosophic subsets definition for dermoscopic image segmentation, *IEEE Access*, 7, p.151047–151053.

Ashour, A.S. and Y. Guo. (2019). Advanced optimization-based neutrosophic sets for medical image denoising, *Neutrosophic Set in Medical Image Analysis*. Elsevier.

Ashour, A.S., Y. Guo, E. Kucukkulahli, p.Erdogmus, and K. Polat (2018a). A hybrid dermoscopy images segmentation approach based on neutrosophic clustering and histogram estimation, *Applied Soft Computing*, 69, p.426–434.

Ashour, A.S., A.R. Hawas, Y. Guo, and M.A. Wahba. (2018b). A novel optimized neutrosophic k-means using genetic algorithm for skin lesion detection in dermoscopy images, *Signal, Image and Video Processing*, 12, p.1311–1318.

Barata, C., J.S. Marques, and J. Rozeira. (2012). A system for the detection of pigment network in dermoscopy images using directional filters, *IEEE Transactions on Biomedical Engineering*, 59, p.2744–2754.

Barata, C., M. Ruela, M. Francisco, T. Mendonça, and J.S. Marques. (2014). Two systems for the detection of melanomas in dermoscopy images using texture and color features. *IEEE Systems Journal*, 8, p.965–979.

Barcelos, C.A.Z., and V. Pires. (2009). An automatic based nonlinear diffusion equations scheme for skin lesion segmentation, *Applied Mathematics and Computation*, 215, p.251–261.

Betta, G., G. Di Leo, G. Fabbrocini, A. Paolillo, and M. Scalvenzi. (2005). Automated application of the “7-point checklist” diagnosis method for skin lesions: estimation of chromatic and shape parameters. Instrumentation and Measurement Technology Conference, 2005. IMTC 2005, Proceedings of the IEEE, 2005, IEEE, p. 1818–1822.

Castiello, G., G. Castellano, and A. M. Fanelli. (2004). A. M. Neuro-fuzzy analysis of dermatological images. Neural Networks, 2004, Proceedings, 2004 IEEE International Joint Conference on, 2004, IEEE, p. 3247–3252.

Celebi, M.E., H. Iyatomi, and G. Schaefer. (2009). Contrast enhancement in dermoscopy images by maximizing a histogram bimodality measure, Image Processing (ICIP), 2009 16th IEEE International Conference on, 2009, IEEE, p. 2601–2604.

Celebi, M. E., H. Iyatomi, W.V. Stoecker, R.H. Moss, H.S. Rabinovitz, G. Argenziano, AND H.P Soyer. (2008). Automatic detection of blue-white veil and related structures in dermoscopy images, *Computerized Medical Imaging and Graphics*, 32, p.670–677.

Celebi, M.E., H.A. Kingravi, and Y.A. Aslandogan. (2007a). Nonlinear vector filtering for impulsive noise removal from color images, *Journal of Electronic Imaging*, 16, p.033008.

Celebi, M.E., H.A. Kingravi, B. Uddin, H. Iyatomi, Y.A. Aslandogan, W.V. Stoecker, and R.H. Moss. (2007b). A methodological approach to the classification of dermoscopy images, *Computerized Medical Imaging and Graphics*, 31, p.362–373.

Celebi, M.E., Q. Wen, H. Iyatomi, K. Shimizu, H. Zhou, and G. Schaefer. (2015). A state-of-the-art survey on lesion border detection in dermoscopy images, *Dermoscopy Image Analysis*, p.97–129.

Chan, T.F., B.Y. Sandberg, and L.A. Vese. (2000). Active contours without edges for vector-valued images, *Journal of Visual Communication and Image Representation*, 11, p.130–141.

Chang, Y., R.J. Stanley, R.H. Moss, and W. Van Stoecker. (2005). A systematic heuristic approach for feature selection for melanoma discrimination using clinical images, *Skin Research and Technology*, 11, p.165–178.

Claridge, E., p.Hall, M. Keefe, and J. Allen. (1992). Shape analysis for classification of malignant melanoma, *Journal of biomedical engineering*, 14, p.229–234.

Clawson, K.M., p. Morrow, B. Scotney, J, Mckenna, and O. Dolan. (2009). Analysis of pigmented skin lesion border irregularity using the harmonic wavelet transform, Machine Vision and Image Processing Conference, 2009, IMVIP'09, 13th International, 2009. IEEE, p. 18–23.

Cokkinides, V., J. Albano, A. Samuels, M. Ward, and J. Thum. (2010). *American cancer society: cancer facts and figures*, Atlanta: American Cancer Society.

Congdon, p.(2007). *Bayesian statistical modelling*. Hoboken, NJ: John Wiley & Sons.

Delalleau, A., J.-M. Lagarde, and J. George. (2011). An a priori shading correction technique for contact imaging devices. *IEEE Transactions on Image Processing*, 20, p. 2876–2885. @@@

Dey, N., V. Rajinikanth, A.S. Ashour, and J.M.R. Tavares. (2018). Social group optimization supported segmentation and evaluation of skin melanoma images, *Symmetry*, 10, p.51.

Di Leo, G., A. Paolillo, P. Sommella, and G. Fabbrocini. (2010). Automatic diagnosis of melanoma: a software system based on the 7-point check-list, System Sciences (HICSS), 2010 43rd Hawaii International Conference on, 2010. IEEE, p. 1–10.

Emre Celebi, M., H.A. Kingravi, H. Iyatomi, Y. Alp Aslandogan, W.V. Stoecker, R.H. Moss, J.M. Malters, et al. (2008). Border detection in dermoscopy images using statistical region merging, *Skin Research and Technology*, 14, p.347–353.

Fawcett, T. (2004). ROC graphs: notes and practical considerations for researchers, *Machine Learning*, 31, p.1–38.

Fleming, M.G., C. Steger, J. Zhang, J. Gao, A.B. Cognetta, and C.R. Dyer. (1998). Techniques for a structural analysis of dermatoscopic imagery, *Computerized Medical Imaging and Graphics*, 22, p.375–389.

Fujita, H., Y. Uchiyama, T. Nakagawa, D. Fukuoka, Y. Hatanaka, T. Hara, T., G.N. Lee, et al. (2008). Computer-aided diagnosis: the emerging of three CAD systems induced by Japanese health care needs, *Computer Methods and Programs in Biomedicine*, 92, p.238–248.

Ganster, H., p.Pinz, R. Rohrer, E. Wildling, M. Binder, and H. Kittler. (2001). Automated melanoma recognition, *IEEE transactions on medical imaging*, 20, p.233–239.

Gao, J., J. Zhang, M.G. Fleming, I. Pollak, and A.B. Cognetta. (1998). Segmentation of dermatoscopic images by stabilized inverse diffusion equations, Image Processing, 1998, ICIP 98, 1998 International Conference on,. IEEE, p. 823–827.

Garnavi, R., M. Aldeen, and J. Bailey. (2012). Computer-aided diagnosis of melanoma using border-and wavelet-based texture analysis, *IEEE Transactions on Information Technology in Biomedicine*, 16, p.1239–1252.

Garnavi, R., M. Aldeen, M.E. Celebi, G. Varigos, and S. Finch. (2011). Border detection in dermoscopy images using hybrid thresholding on optimized color channels, *Computerized Medical Imaging and Graphics*, 35, p.105–115.

Gómez, D.D., C. Butakoff, B.K. Ersboll, and W. Stoecker, W. 2008. Independent histogram pursuit for segmentation of skin lesions, *IEEE Transactions on Biomedical Engineering*, 55, p.157–161.

Guo, Y., A.S. Ashour, and F. Smarandache. (2018). A novel skin lesion detection approach using neutrosophic clustering and adaptive region growing in dermoscopy images, *Symmetry*, 10, p.119.

Han, J., J. Pei, and M. Kamber. (2011). *Data mining: concepts and techniques, Elsevier*.

Handels, H., T. Roß, J. Kreusch, H.H. Wolff, and S.J. Poeppl. (1999). Feature selection for optimized skin tumor recognition using genetic algorithms, *Artificial Intelligence in Medicine*, 16, p.283–297.

Hawas, A.R., Y. Guo, C. Du, K. Polat, and A.S. Ashour. (2019). OCE-NGC: A neutrosophic graph cut algorithm using optimized clustering estimation algorithm for dermoscopic skin lesion segmentation, *Applied Soft Computing*, p.105931.

Haykin, S. (1994). *Neural networks, a comprehensive foundation*. Macmilan.

Hibler, B. P., Q. Qi, and A.M. Rossi. (2016). Current state of imaging in dermatology, Seminars in cutaneous medicine and surgery, *Frontline Medical Communications*, p.2–8.

Iyatomi, H., M.E. Celebi, G. Schaefer, and M. Tanaka. (2011). Automated color calibration method for dermoscopy images, *Computerized Medical Imaging and Graphics*, 35, p.89–98.

Iyatomi, H., K.-A. Norton, M.E. Celebi, G. Schaefer, M. Tanaka, and K. Ogawa. (2010). Classification of melanocytic skin lesions from non-melanocytic lesions, Engineering in Medicine and Biology Society (EMBC), 2010 Annual International Conference of the IEEE, IEEE, p. 5407–5410.

Iyatomi, H., H. Oka, M.E. Celebi, M. Hashimoto, M. Hagiwara, M. Tanaka, and k. Ogawa. (2008a). An improved internet-based melanoma screening system with dermatologist-like tumor area extraction algorithm, *Computerized Medical Imaging and Graphics*, 32, p.566–579.

Iyatomi, H., H. Oka, M.E. Celebi, K. Ogawa, G. Argenziano, H.P/ Soyer, H. Koga, et al. (2008b). Computer-based classification of dermoscopy images of melanocytic lesions on acral volar skin, *Journal of Investigative Dermatology*, 128, p.2049–2054.

Jafari-Khouzani, K., and H. Soltanian-Zadeh. (2005). Radon transform orientation estimation for rotation invariant texture analysis. *IEEE Transactions on Pattern Analysis and Machine Intelligence*, 27, p.1004–1008.

Jianli, L., and Z. Baoqi. (2009). The segmentation of skin cancer image based on genetic neural network, Computer Science and Information Engineering, 2009 WRI World Congress on IEEE, p. 594–599.

Johr, R.H. (2002). Dermoscopy: alternative melanocytic algorithms—the ABCD rule of dermatoscopy, menzies scoring method, and 7-point checklist, *Clinics in Dermatology*, 20, p.240–247.

Kawahara, J., S. Daneshvar, G. Argenziano, and G. Hamarneh. (2018). Seven-point checklist and skin lesion classification using multitask multimodal neural nets, *IEEE Journal of Biomedical and Health Informatics*, 23, p.538–546.

Khan, A., K. Gupta, R.J. Stanley, W.V. Stoecker, R.H. Moss, G. Argenziano, H.P. Soyer, et al. (2009). Fuzzy logic techniques for blotch feature evaluation in dermoscopy images, *Computerized Medical Imaging and Graphics*, 33, p.50–57.

Lee, T., V. Ng, R. Gallagher, A. Coldman, and D. Mclean. (1997). Dullrazor®: A software approach to hair removal from images, *Computers in Biology and Medicine*, 27, p.533–543.

Lee, T.K., D.I. Mclean, and M.S. Atkins. (2003). Irregularity index: a new border irregularity measure for cutaneous melanocytic lesions, *Medical Image Analysis*, 7, p.47–64.

Lissner, I., and p.Urban. (2012). Toward a unified color space for perception-based image processing. *IEEE Transactions on Image Processing*, 21, p.1153–1168.

Ma, Z., and J.M.R. Tavares (2016). A novel approach to segment skin lesions in dermoscopic images based on a deformable model, *IEEE Journal of Biomedical and Health Informatics*, 20, p.615–623.

Maeda, J., A. Kawano, S. Saga, and Y. Suzuki. (2007). Number-driven perceptual segmentation of natural color images for easy decision of optimal result, Image Processing, 2007. ICIP 2007, IEEE International Conference on IEEE, II-265-II-268.

Maeda, J., A. Kawano, S. Yamauchi, Y. Suzuki, A. Marçal, and T. Mendonça. (2008). Perceptual image segmentation using fuzzy-based hierarchical algorithm and its application to dermoscopy images, Soft Computing in Industrial Applications, SMCia'08, IEEE Conference on, 2008, IEEE, p. 66–71.

Maglogiannis, I., and C.N. Doukas. (2009). Overview of advanced computer vision systems for skin lesions characterization, *IEEE Transactions on Information Technology in Biomedicine*, 13, p.721–733.

Maglogiannis, I., E. Zafiropoulos, and C. Kyranoudis. (2006). Intelligent segmentation and classification of pigmented skin lesions in dermatological images. Hellenic Conference on Artificial Intelligence, 2006. Springer, p. 214–223.

Masood, A., and A.A. Al-Jumaily. (2013). Fuzzy C mean thresholding based level set for automated segmentation of skin lesions, *Journal of Signal and Information Processing*, 4, p.66.

Masood, A., and A.A. Al-Jumaily. (2013). Computer aided diagnostic support system for skin cancer: a review of techniques and algorithms, *International Journal of Biomedical Imaging*, 2013.

Mayer, J.E., and R.H. Goldman. (2016). Arsenic and skin cancer in the USA: the current evidence regarding arsenic-contaminated drinking water, *International Journal of Dermatology*, 55.

Mumford, D., and J. Shah. (1989). Optimal approximations by piecewise smooth functions and associated variational problems, *Communications on Pure and Applied Mathematics*, 42, p.577–685.

Nagieb, R.M., A.S. Ashour, Y. Guo, H.A. El-Khobby, M.M.A. and Elnaby. (2018). Initialization of active contour for dermoscopic image segmentation: a comparative study,2018 IEEE

International Symposium on Signal Processing and Information Technology (ISSPIT), IEEE, p. 370–374.

Narayanan, D.L., R.N. Saladi, and J.L. Fox. (2010). Ultraviolet radiation and skin cancer, *International Journal of Dermatology*, 49, p.978–986.

Ng, V.T., B.Y. Fung, and T.K. Lee. (2005). Determining the asymmetry of skin lesion with fuzzy borders, *Computers in Biology and Medicine*, 35, p.103–120.

Nie, D. (2011). Classification of melanoma and clark nevus skin lesions based on Medical Image Processing Techniques, Computer Research and Development (ICCRD), 2011 3rd International Conference on IEEE, p. 31–34.

Nock, R., and F. Nielsen. (2004). Statistical region merging, *IEEE Transactions on Pattern Analysis and Machine Intelligence*, 26, p.1452–1458.

Norton, K.A., H. Iyatomi, M.E. Celebi, S. Ishizaki, M. Sawada, R. Suzaki, K. Kobayashi, et al. (2012). Three-phase general border detection method for dermoscopy images using non-uniform illumination correction, *Skin Research and Technology*, 18, p.290–300.

Oliveira, R.B., E. Mercedes Filho, Z. Ma, J.P. Papa, A.S. Pereira, and J.M.R. Tavares. (2016). Computational methods for the image segmentation of pigmented skin lesions: a review, *Computer Methods and Programs in Biomedicine*, 131, p.127–141.

Otsu, N. (1979). A threshold selection method from gray-level histograms, *IEEE Transactions on Systems, Man, and Cybernetics*, 9, p.62–66.

Pires, V.B., and C.A.Z. Barcelos. (2007). Edge detection of skin lesions using anisotropic diffusion, Intelligent Systems Design and Applications, 2007, ISDA 2007, Seventh International Conference on, 2007, IEEE, p. 363–370.

Pizer, S.M., E.P. Amburn, J.D. Austin, R. Cromartie, A. Geselowitz, T. Greer, B. Ter Haar Romeny, et al. (1987). Adaptive histogram equalization and its variations, *Computer Vision, Graphics, and Image Processing*, 39, p.355–368.

Rahman, M.M., p.Bhattacharya, and B.C. Desai. (2008). A multiple expert-based melanoma recognition system for dermoscopic images of pigmented skin lesions, BioInformatics and BioEngineering, 2008, BIBE 2008, 8th IEEE International Conference on IEEE, p. 1–6.

Ramlakhan, K., and Y. Shang. (2011). A mobile automated skin lesion classification system, Tools with Artificial Intelligence (ICTAI), 2011 23rd IEEE International Conference on IEEE, p. 138–141.

Rohrer, R., H. Ganster, A. Pinz, and M. Binder. (1998). Feature selection in melanoma recognition. Pattern Recognition, 1998, Proceedings, Fourteenth International Conference on IEEE, p.1668–1670.

Roß, T., H. Handels, J. Kreusch, H. Busche, H. Wolf, and S.J. Pöppl. (1995). Automatic classification of skin tumours with high resolution surface profiles, International Conference on Computer Analysis of Images and Patterns, 1995. Springer, p. 368–375.

Ruiz, D., V. Berenguer, A. Soriano, and B. Sánchez. (2011). A decision support system for the diagnosis of melanoma: a comparative approach, *Expert Systems with Applications*, 38, p.15217–15223.

Ruocco, E., G. Argenziano, G. Pellacani, and S. Seidenari. (2004). Noninvasive imaging of skin tumors, *Dermatologic Surgery*, 30, p.301–310.

Schmid-Saugeona, P., J. Guillodb, and J.-P. Thirana. (2003). Towards a computer-aided diagnosis system for pigmented skin lesions, *Computerized Medical Imaging and Graphics*, 27, p.65–78.

She, Z., Y. Liu, and A. Damatoa. (2007). Combination of features from skin pattern and ABCD analysis for lesion classification, *Skin Research and Technology*, 13, p.25–33.

Sheha, M.A., M.S. Mabrouk, and A. Sharawy (2012). Automatic detection of melanoma skin cancer using texture analysis, *International Journal of Computer Applications*, 42, p.22–26.

Shimizu, K., H. Iyatomi, M.E. Celebi, K.-A. Norton, and M. Tanaka (2015). Four-class classification of skin lesions with task decomposition strategy, *IEEE Transactions on Biomedical Engineering*, 62, p.274–283.

Silva, C.S., and A.R. Marcal. (2013). Colour-based dermoscopy classification of cutaneous lesions: an alternative approach, *Computer Methods in Biomechanics and Biomedical Engineering: Imaging & Visualization*, 1, p.211–224.

Silveira, M., J.C. Nascimento, J.S. Marques, A.R. Marçal, T. Mendonça, S. Yamauchi, J. Maeda, et al. (2009). Comparison of segmentation methods for melanoma diagnosis in dermoscopy images, *IEEE Journal of Selected Topics in Signal Processing*, 3, p.35–45.

Smith, L., and S. Macneil. (2011). State of the art in non-invasive imaging of cutaneous melanoma, *Skin Research and Technology*, 17, p.257–269.

Steinwart, I., and A. Christmann. (2008). *Support vector machines*, Springer Science & Business Media.

Stockham, T.G. (1972). Image processing in the context of a visual model, *Proceedings of the IEEE*, 60, p.828–842.

Surowka, G. (2008). Supervised learning of melanocytic skin lesion images, Human System Interactions, 2008 Conference on IEEE, p. 121–125.

Vennila, G.S., L.P. Suresh, and K. Shunmuganathan. (2012). Dermoscopic image segmentation and classification using machine learning algorithms, Computing, Electronics and Electrical Technologies (ICCEET), 2012 International Conference on, 2012. IEEE, p. 1122–1127.

Wahba, M.A., A.S. Ashour, Y. Guo, S.A. Napoleon, and M.M.A. Elnaby. (2018). A novel cumulative level difference mean based GLDM and modified ABCD features ranked using eigenvector centrality approach for four skin lesion types classification, *Computer Methods and Programs in Biomedicine*, 165, p.163–174.

Wahba, M.A.,A.S. Ashour, S.A. Napoleon, M.M.A. Elnaby, and Guo, Y. (2017). Combined empirical mode decomposition and texture features for skin lesion classification using quadratic support vector machine, *Health information science and systems*, 5, p.10.

Wassef, C., and B.K. Rao. (2013). Uses of non-invasive imaging in the diagnosis of skin cancer: an overview of the currently available modalities, *International journal of dermatology*, 52, p.1481–1489.

Wighton, P., T.K. Lee, H. Lui, D. Mclean, and M.S. Atkins. (2011). Chromatic aberration correction: an enhancement to the calibration of low-cost digital dermoscopes, *Skin Research and Technology*, 17, p.339–347.

Wong, A., and J. Scharcanski, and p.Fieguth. (2011). Automatic skin lesion segmentation via iterative stochastic region merging, *IEEE Transactions on Information Technology in Biomedicine*, 15, p.929–936.

Xie, F.-Y., S.-Y. Qin, Z.-G. Jiang, and R.-S. Meng. (2009). PDE-based unsupervised repair of hair-occluded information in dermoscopy images of melanoma, *Computerized Medical Imaging and Graphics*, 33, p.275–282.

Yuan, X., Z. Yang, G. Zouridakis, and N, Mullani. (2006). SVM-based texture classification and application to early melanoma detection, Engineering in Medicine and Biology Society, 2006, EMBS'06. 28th Annual International Conference of the IEEE. IEEE, p.4775–4778.

Yüksel, M.E., and M. Borlu. (2009). Accurate segmentation of dermoscopic images by image thresholding based on type-2 fuzzy logic, *IEEE Transactions on Fuzzy Systems*, 17, p.976–982.

Zhou, H., M. Chen, R. Gass, J.M. Rehg, L. Ferris, J. Ho, and L. Drogowski. (2008). Feature-preserving artifact removal from dermoscopy images, Medical Imaging 2008: Image Processing, 2008, International Society for Optics and Photonics, p. 69141B.

Zhou, H., X. Li, G. Schaefer, M.E. Celebi, and p.Miller. (2013). Mean shift based gradient vector flow for image segmentation, *Computer Vision and Image Understanding*, 117, p.1004–1016.

Zhou, H., G. Schaefer, M.E. Celebi, H. Iyatomi, A.-K. Norton, T. Liu, and F. Lin. (2011). Skin lesion segmentation using an improved snake model, Engineering in Medicine and Biology Society (EMBC), 2010 Annual International Conference of the IEEE. IEEE, p. 1974–1977.

Conclusions

Engineering and technology have the potential to revolutionise patient care and health outcomes. This book was developed so that readers can:

- Recognize the factors that continue to drive change in healthcare
- Discuss the potential for development of new medical devices, tools and process that can benefit society.

Despite the positive potential that engineering and technology can bring to the healthcare profession, there is evidence of much waste in healthcare research. In the second chapter, we showed that there is a tendency for research to be "technology led" rather than "needs led" and evidence that funding for enterprising firms in health technology is not always aligned with health research funding streams. Therefore, HTA is required to inform a decision in the development of a potential technology.

Moreover, we demonstrated how developments in sensor technologies have enabled medical professionals to accurately diagnose diseases. These technologies aim to prevent the disease by detecting precursor symptoms to a condition, rather than just reacting and detecting problems as they occur. Moreover, there is interest in shifting patient care services from specialised centres to individual home environments. Therefore, we explained how radar will become "a sensor in a suite of sensors" within the ambient space of a patient's home. These sensors will monitor temperature and humidity, as well as other sensors from mobile and wirelessly connected IoT devices. Since these devices will provide different information at different data rates, their information will need to be properly fused in order to increase the accuracy of prediction models. This paradigm shift is vital for the healthcare system and its sustainability in the future.

Moreover, with the advent of 5G and beyond, our living environments will become more connected than ever. 5G is foreseen as a key enabler for indoor sensing networks with device-to-device communication and mobile edge computing. In fact, 5G will also be an enabler for applications such as remote surgery, e-health, and ambient patient monitoring. These will be supported by artificial intelligence, cloud/cognitive computing, commoditisation of data from crowdsourcing, and next-generation communication.

Machine learning algorithms will also play an important role in healthcare applications. For example, these algorithms will be used to accurately detect cancers. They will also need to become adaptive to take into account sensors joining and leaving the ambient IoT environment. Furthermore, this adaptive capability will need to deal with sensor malfunctions,

Engineering and Technology for Healthcare, First Edition.
Edited by Muhammad Ali Imran, Rami Ghannam and Qammer H. Abbasi.

sensors working in a synchronous or asynchronous manner, and sensors being moved or wearables being worn or not. All these phenomena change the way algorithms need to be designed in order to predict a class of actions, or estimate vital signs that can infer information for prevention.

Moreover, we discussed recent advances in implantable devices for detecting and diagnosing neurological diseases. We have demonstrated how these technologies can intimately be integrated with the soft, curvilinear tissues of the body in ways that are impossible with conventional, wafer-based forms of electronics. Although the functions achieved thus far rely on diverse, physical modes of interaction with tissue, future neural devices might also incorporate MRI compatibility, long-term neural modulation capability for which new sensors and microelectromechanical and microfluidic components in flexible/stretchable formats will be required. Innovative ideas in materials and device engineering will be required, as will an improved fundamental understanding of the physical chemistry of the biotic/abiotic interface. Other areas for research include development of components for mechanical, thermal, or chemical energy scavenging and of systems for wireless communications. These engineering goals, the foundational science that underpins them, and their relevance to improvements in human health will drive interest in this emerging field for many years to come.

Next, we demonstrated the different energy harvesting and scavenging techniques that can be used to ensure that these implantable and wearable electronic products are sustained. We discussed the main techniques used in the literature and critiqued each. We mentioned that using a hybrid approach that combines multiple energy harvesters on the same platform is most likely the best solution.

Similarly, we discussed recent progress in wireless communication and control systems. Promising applications such as remote surgery and remote diagnosis require careful communication and control co-design. System models for co-design approaches were explained, and simulation results were presented to highlight the benefits of co-design in terms of communication and control performance.

In summary, the major goal of the engineering profession is to serve society. Maintaining and improving people's health requires scientific breakthroughs and technological developments, as well as governmental policies and initiatives. In this book, we aimed to offer a set of resources and approaches that can be adapted and implemented. We hope that these resources will continue to grow as others contribute. We also hope that others can design and develop the products, processes and systems that enable us to lead a healthier future.

Index

Engineering and Technology for Healthcare, First Edition.
Edited by Muhammad Ali Imran, Rami Ghannam and Qammer H. Abbasi.

w

y

z